DESIGNERS' GUIDES TO THE EUROCODES

DESIGNERS' GUIDE TO EN 1991-1-4 EUROCODE 1: ACTIONS ON STRUCTURES, GENERAL ACTIONS

PART 1–4: WIND ACTIONS

Eurocode Designers' Guide Series

Designers' Guide to EN 1990. Eurocode: Basis of Structural Design. H. Gulvanessian, J.-A. Calgaro and M. Holický. 0 7277 3011 8. Published 2002.

Designers' Guide to EN 1994-1-1. Eurocode 4: Design of Composite Steel and Concrete Structures. Part 1.1: General Rules and Rules for Buildings. R. P. Johnson and D. Anderson. 0 7277 3151 3. Published 2004.

Designers' Guide to EN 1997-1. Eurocode 7: Geotechnical Design – General Rules. R. Frank, C. Bauduin, R. Driscoll, M. Kavvadas, N. Krebs Ovesen, T. Orr and B. Schuppener. 0 7277 3154 8. Published 2004.

Designers' Guide to EN 1993-1-1. Eurocode 3: Design of Steel Structures. General Rules and Rules for Buildings. L. Gardner and D. Nethercot. 0 7277 3163 7. Published 2004. *but does this include Annexe for towers and masts? Pt. 3.1? 2006?*

Designers' Guide to EN 1992-1-1 and EN 1992-1-2. Eurocode 2: Design of Concrete Structures. General Rules and Rules for Buildings and Structural Fire Design. A. W. Beeby and R. S. Narayanan. 0 7277 3105 X. Published 2005.

— check from (red) copy
— instead 1993-3-1

Designers' Guide to EN 1998-1 and EN 1998-5. Eurocode 8: Design of Structures for Earthquake Resistance. General Rules, Seismic Actions, Design Rules for Buildings, Foundations and Retaining Structures. M. Fardis, E. Carvalho, A. Elnashai, E. Faccioli, P. Pinto and A. Plumier. 0 7277 3348 6. Published 2005.

Designers' Guide to EN 1995-1-1. Eurocode 5: Design of Timber Structures. Common Rules and for Rules and Buildings. C. Mettem. 0 7277 3162 9. Forthcoming: 2007 (provisional).

Designers' Guide to EN 1991-4. Eurocode 1: Actions on Structures. Wind Actions. N. Cook. 0 7277 3152 1. Forthcoming: 2007 (provisional). *→this guide!*

Designers' Guide to EN 1996. Eurocode 6: Part 1.1: Design of Masonry Structures. J. Morton. 0 7277 3155 6. Forthcoming: 2007 (provisional).

Designers' Guide to EN 1991-1-2, 1992-1-2, 1993-1-2 and EN 1994-1-2. Eurocode 1: Actions on Structures. Eurocode 3: Design of Steel Structures. Eurocode 4: Design of Composite Steel and Concrete Structures. Fire Engineering (Actions on Steel and Composite Structures). Y. Wang, C. Bailey, T. Lennon and D. Moore. 0 7277 3157 2. Forthcoming: 2007 (provisional).

Designers' Guide to EN 1993-2. Eurocode 3: Design of Steel Structures. Bridges. C. R. Hendy and C. J. Murphy. 0 7277 3160 2. Forthcoming: 2007 (provisional).

Designers' Guide to EN 1991-2, 1991-1-1, 1991-1-3 and 1991-1-5 to 1-7. Eurocode 1: Actions on Structures. Traffic Loads and Other Actions on Bridges. J.-A. Calgaro, M. Tschumi, H. Gulvanessian and N. Shetty. 0 7277 3156 4. Forthcoming: 2007 (provisional).

Designers' Guide to EN 1991-1-1, EN 1991-1-3 and 1991-1-5 to 1-7. Eurocode 1: Actions on Structures. General Rules and Actions on Buildings (not Wind). H. Gulvanessian, J.-A. Calgaro, P. Formichi and G. Harding. 0 7277 3158 0. Forthcoming: 2007 (provisional).

Designers' Guide to EN 1994-2. Eurocode 4: Design of Composite Steel and Concrete Structures. Part 2: General Rules and Rules for Bridges. C. R. Hendy and R. P. Johnson. 07277 3161 0. Published 2006.

www.eurocodes.co.uk

DESIGNERS' GUIDES TO THE EUROCODES

DESIGNERS' GUIDE TO EN 1991-1-4
EUROCODE 1: ACTIONS ON STRUCTURES, GENERAL ACTIONS

PART 1–4: WIND ACTIONS

N. COOK

Published by Thomas Telford Publishing, Thomas Telford Ltd, 1 Heron Quay, London E14 4JD
URL: http://www.thomastelford.com

Distributors for Thomas Telford books are
USA: ASCE Press, 1801 Alexander Bell Drive, Reston, VA 20191-4400
Japan: Maruzen Co. Ltd, Book Department, 3–10 Nihonbashi 2-chome, Chuo-ku, Tokyo 103
Australia: DA Books and Journals, 648 Whitehorse Road, Mitcham 3132, Victoria

First published 2007

Eurocodes Expert

Structural Eurocodes offer the opportunity of harmonized design standards for the European construction market and the rest of the world. To achieve this, the construction industry needs to become acquainted with the Eurocodes so that the maximum advantage can be taken of these opportunities

Eurocodes Expert is a new ICE and Thomas Telford initiative set up to assist in creating a greater awareness of the impact and implementation of the Eurocodes within the UK construction industry

Eurocodes Expert provides a range of products and services to aid and support the transition to Eurocodes. For comprehensive and useful information on the adoption of the Eurocodes and their implementation process please visit our website or email eurocodes@thomastelford.com

---

A catalogue record for this book is available from the British Library

ISBN: 978-0-7277-3152-4

Typeset by Academic + Technical, Bristol
Printed and bound in Great Britain by MPG Books, Bodmin

Preface

EN 1991-1-4, *Eurocode 1: Actions on Structures – General Actions – Part 1–4: Wind Actions*, is the head code for wind actions on structures and describes the principles and requirements for calculating design wind loads on structures. It complies with the requirements of Eurocode EN 1990, *Eurocode: Basis of Structural Design*, and provides the wind actions necessary to implement the structural design Eurocodes 2 to 9.

Aims and objectives of this guide

The principal aim of this book is to provide the user with guidance on the interpretation and use of EN 1991, *Eurocode 1: Actions on Structures – General Actions – Part 1–4: Wind Actions*, and on the expected changes that will be introduced by the associated National Annexes. In particular, the opportunity has been taken to add a commentary on the changes introduced in the UK National Annex (Draft for public comment). Some of the remaining issues with the UK NAD may be resolved before publication.

Layout of this guide

EN 1991-1-4 has a foreword and eight sections together with six annexes. Chapters 1 to 8 of this guide correspond to the eight sections of EN 1991-1-4. The numbering sequence of these chapters mirrors the clause numbers of the EN. Chapter 9 corresponds to the six annexes. In addition, the relevant clause numbers of EN 1991-1-4 and of the UK National Annex appear in the margins against the corresponding commentary.

Text which has been directly reproduced from the EN is set in italics. Expressions and figures repeated from EN 1994-1-4 retain their number. The author's expressions, where numbered and figures have numbers prefixed by D (for Designers' Guide); for example, equation (D4.1) in Chapter 4.

Acknowledgements

This guide would not have been possible without the analysis and calibrations of EN 1991-1-4 undertaken by the other members of the UK National Application Document drafting panel, who have given their time and expertise freely to assist the British Standards Institution to prepare for the implementation of the Eurocode in the UK.

Contents

CHAPTER 1

General

This chapter is concerned with the general aspects of EN 1991-1-4, *Eurocode: Actions on Structures – General Actions – Part 1–4: Wind Actions*. The material in this chapter is covered in *Section 1*, in the following clauses:

1.1. Scope

EN 1991-1-4, *Eurocode: Actions on Structures – General Actions – Part 1–4: Wind Actions* is meant to give the characteristic values of wind actions on all land-based structures, to encompass the whole structure, parts of the structure (e.g. walls, roofs) and elements attached to the structure (e.g. chimneys, canopies). Previously in the UK, the characteristic wind loads for various types of structure were either:

- embodied in the relevant standards for that structure (e.g. BS 5400 for bridges, BS 6399-2[1] for buildings and BS 8100 for lattice towers and masts), or
- the standard referred the designer to a 'head code' – usually BS 6399-2[1] – for the characteristic wind loads and defined how these loads should be applied to the component (e.g. BS 5534 for slating and tiling).

EN 1991-1-4 is therefore intended as the source head code for all land-based structures. However, this is almost impossible to achieve because each different type of structure requires a different kind of information about the wind loads. Designing bridges requires information on how the wind loads vary horizontally along the length of the bridge, while designing masts requires information on how wind loads vary vertically up the height of the mast. It is necessary to know the frequency spectrum of the gust loads when the structure is dynamic, but only the characteristic largest load is required where a static design is appropriate.

Clause 1.1(3)

However, this first version of EN 1991-1-4 covers only buildings up to 200 m in height and bridges of span less than 200 m. It excludes torsional vibrations of buildings, vibrations of bridges due to transverse (crosswind) turbulence, cable-supported bridges and vibrations of a structure in modes other than the fundamental mode. The note to clause 1.1(12) states: '*The National Annex may provide guidance on these aspects as non-contradictory complementary information*' but, in reality, the CEN drafting rules for the National Annex

Clause 1.1(2)
Clause 1.1(11)

Clause 1.1(12)
UK NA 2.1

(NA) do not permit the introduction of any new information – only **references** to non-contradictory complementary information (NCCI). (See the discussion on the distinction between Principles and Application Rules in section 1.4, below, for a fuller description of the restrictions on the content of the NA and the meaning of the term 'non-contradictory complementary information'.) Wind loads on guyed masts and lattice towers are covered separately in EN 1993-3-1 and lighting columns in EN 40. The UK NA clause 2.1 refers to a background document[xx] and two external references[15,yy].

*xx = PD 6688
(in preparation)
15 = Guide by N Cook
yy = ASCE Manual 67
(wind tunnel studies)*

The initial version of EN 1991-1-4 cannot be implemented without the accompanying NA issued by the national authority for the member state where the structure is to be built. The NA provides data and guidance that is geographically dependent, such as the design wind speed map, national choices for the National Values, nationally determined parameters (NDPs), and other choices where a common approach has not been agreed, as well as rules necessary to meet local building regulations and laws. The EN and accompanying NA will exist in parallel with the pre-existing national code (BS 6399-2[1] in the UK) for the 'co-existence period', expected to be about five years. It is intended that the national codes will be withdrawn and that these NAs will be incorporated into the main body of the EN at the end of this period to complete the 'harmonization' process, provided an accommodation can be reached on the outstanding issues of disagreement. In the mean time, the user of EN 1991-1-4 is required to address the additions and exceptions defined in the NA for the member state **where the structure is to be built** – not the NA for the Member State where the designer is resident.

In effect, the combined EN 1991-1-2 and the relevant NA becomes a 'national' code which will be different for each Member State – albeit with a common approach and structure imposed by the EN. Nevertheless, the net result is a long way from providing 'harmonized' rules applicable across all Member States. This was inevitable, given that the strength of the wind varies greatly across Europe, so that the importance of wind loading in comparison with other actions also varies widely, and this is reflected in different regulatory frameworks and design practices in each Member State.

Clearly, it is not possible for this Guide to address all the National Annexes because of the large number of NAs and because none were completed at the time of writing. However, it has been possible to address the UK NA and to use this as an example of how an NA will modify and augment the provisions of the EN. In particular, this Guide will illustrate some fundamental differences in approach between the EN and the UK NA, which highlight the difficulty of achieving a common approach across member states with widely differing current practice.

Two fundamental differences will quickly become obvious:

- The EN simplifies implementation by adopting simplifying approximations which introduce unnecessary errors. Although some of these errors compensate for others, the errors are generally conservative. Nevertheless, the EN methodology includes equations that are quite complex and, therefore, prone to calculation errors.
- The UK NA does not make simplifying assumptions but implements the best current practice using factors derived from look-up tables and graphs, requiring only simple addition and multiplication to accumulate the required results.

Clause 1.1(4)

As the base meteorological information for any location is strongly dependent on the local climate and regional geography, the EN requires this to be given in the NA, or through National Values in the main text. The meteorological data in the UK NA are restricted to fully developed storm systems and exclude local thermal effects, such as tornadoes and

Clause 1.1(10)

'funnelling' – which here refers to funnelling of the wind into valleys, not funnelling between adjacent buildings. Other NAs are expected to include local effects where these are critical to the design, especially in mountainous regions subject to katabatic (gravity) winds.

The EN provides certain necessary information in the form of **informative** annexes:

- Annex A gives illustrations of terrain categories, rules for transitions between roughness categories, rules of orography (hills, escarpments, etc.) and effects of upwind buildings. *Clause 1.1(5)*
- Annex B, C and D give two alternative calculation procedures and a graphical method for some types of structure, respectively, for determining the structural factor $c_s c_d$, the factor that describes the effects of structural size and dynamics on the wind actions. *Clause 1.1(6)* *Clause 1.1(7)*
- Annex E gives rules for vortex-induced response, including two alternative calculation procedures, and guidance on other aeroelastic effects. *Clause 1.1(8)*
- Annex F gives guidance on the dynamic characteristics of linear structures – fundamental natural frequencies, mode shapes and damping. *Clause 1.1(9)*

The **informative** status of these annexes and the need to include alternative procedures reflect the inability of the Member States to agree on quite fundamental procedures. The CEN drafting rules require the national authority to adopt or reject as **normative** any Annex in its entirety. It is not permitted to adopt some parts of an Annex or to 'cherry-pick' individual clauses – it is a case of all or nothing. However, it should be expected that national authorities will take advantage of the extensive provisions for national choice in the main text to circumvent this restriction, i.e. the national authorities may adopt the recommended values for the National Values or supply their own, but in the extreme case the national authority might rewrite an entire Annex in order to change some parts of it.

The CEN rules also prevent the national authority from amalgamating the EN and NA into a single coherent document for use in the Member State. However sensible a solution this may seem to be, CEN rules appear to be driven more by the politics of 'harmonization' than the practicalities of implementation. It will therefore be interesting to observe which Member States adhere to this rule. In default of a single document, the single most useful action a user can do is to annotate the EN by writing the NA clause number in the margin of the clause in the EN which is modified or overridden by the NA. This will complement the EN clause references in the NA, providing complete cross-references between the two documents.

As stated earlier, the EN with the relevant NA produces a different 'national' code for each Member State. **Accordingly, users of EN 1991-1-4 must at all times be aware of the exceptions or additions to the EN given in the National Annex for the appropriate Member State.** *Important warning*

1.2. Normative references

EN 1991-1-4 refers to some provisions that are specified in other Eurocodes, specifically: *Clause 1.2*

EN 1990	*Eurocode: basis of structural design*
EN 1991-1-3	*Eurocode 1: Actions on structures: Parts 1–3: Snow actions*
EN 1991-1-6	*Eurocode 1: Actions on structures: Parts 1–6: Actions during execution*
EN 1991-2	*Eurocode 1: Actions on structures: Part 2: Traffic actions on bridges*

Where the reference in the text is dated, the designer **must** refer to that specified edition. Where the reference is undated, the designer **must** refer to the latest edition.

Effectively, this clause requires the designer to have access to copies of all the relevant structural Eurocodes, the majority of which may be irrelevant to most designs. Indeed, it may be necessary to have both the current and a previous dated copy of an EN – for example, clause 1.3(1)P refers to EN 1990 (undated, hence current version), while clause 1.4(1)P refers specifically to EN 1990, 2002. In addition, each Eurocode may require a National Annex which will be different for every member state. This may be anathema to designers who have been used to finding all the information required for a structural form in a single self-contained document, e.g. a British Standard, such as BS 5400.

1.3. Assumptions

The assumptions of EN 1990, clause 1.3, are applicable to EN 1991-1-4. These are: *Clause 1.3(1)P*

'the choice of the structural system and the design of a structure is made by appropriately qualified and experienced personnel';
'execution is carried out by personnel having the appropriate skill and experience';
'adequate supervision and quality control is provided during execution of the work, i.e. in design offices, factories, plants and on site';
'the construction materials and products are used as specified in EN 1990 or in EN 1991 to EN 1999 or in the relevant execution standards, or reference material or product specifications';
'the structure will be adequately maintained';
'the structure will be used in accordance with the design assumptions'.

The implications of these assumptions are discussed in the *Designers' Guide to EN 1990. Eurocode: Basis of Structural Design.*[2] However, we may note here that the eventual owner or user of a building designed using EN 1991-1-4 must be made aware of his/her responsibilities to implement a maintenance scheme and to ensure that overloading does not take place. As a practical example, where the design assumes that an elective dominant opening remains closed at the ultimate or service limit, e.g. the main doors of an aircraft hangar, the responsibility to ensure that the opening is properly closed in strong winds devolves to the owner/user. In the UK, this places a new responsibility on the owner/user of the building.

1.4. Distinction between Principles and Application Rules
The rules in EN 1990, 2002, clause 1.4 apply:

Clause 1.4(1)P

'The Principles comprise: general statements for which there are no alternative, as well as: requirements and analytical models for which no alternative is permitted unless specifically stated.'
'The Principles are identified by the letter 'P' following the paragraph number.'
'The Application Rules are generally recognized rules which follow the principles and satisfy their requirements.'
'It is permissible to use alternative design rules different from the Application Rules given in EN 1990 provided that it can be demonstrated that they comply with the principles and are at least equivalent with regard to structural safety, serviceability and durability which would be expected when using the Eurocodes.'
    'Note: If an alternative rule is substituted for an application rule, the resulting design cannot be claimed to be wholly in accordance with EN 1990 although the design will remain in accordance with the Principles of EN 1990. When EN 1990 is used in respect of a property listed in an Annex Z of a product standard or ETAG, the use of an alternative design rule may not be acceptable for CE marking.'
'In EN 1990, the Application Rules are identified by a number in brackets.'

It is assumed that adoption of this rule implies that each of these references to EN 1990 apply to EN 1991-1-4.

The designer **must** follow the principles, which are clauses marked 'P' and where use of the word 'shall' indicates there is no alternative. The designer need not use the Application Rules, where the use of 'should' or 'may' indicates that there is an alternative. Where the designer chooses to follow other rules, the onus is on him/her to demonstrate equivalence in terms on safety, serviceability and durability. However, in choosing to follow other rules that are wholly in accordance with the principles, the designer cannot claim that the design is wholly in accordance with the EN.

It is not possible for a National Annex to address this problem because the EC guidance paper *Application and Use of Eurocodes*[3] states:

'National Provisions should avoid replacing EN provisions...'
and

'When, however, National Provisions do provide that the design may – even after the end of the co-existence period – deviate from the EN Eurocodes or certain provisions thereof do not apply (e.g. Application rules) then the design will not be called "a design according to the Eurocodes".'

Note that 'should' implies replacement is permissible but may lead to difficulties when the NAs are incorporated into the EN.

So, while the distinction between Principle and Application Rule was originally chosen to encourage innovation in design, the requirements of clause 1.4(1)P apparently conspire to stifle any innovation by the appropriate authority, the client or the individual engineer. However, *Designers' guide to EN 1990*[2] notes that, while use of alternative rules would cause problems in granting a CE marking to a product (e.g. to a standardized structure made in quantity), it is *'workable for an individual design'* (e.g. a unique building).

In any case, the Eurocodes permit the use of *'non-contradictory complementary information'* to supplement the information given within. It is not reasonable to expect every possible shape of structure to be included, so, for example, pressure coefficients derived using the principles for other shapes would be applicable. Therefore, information which is not included in the EN, such as pressure coefficients for an unusual shape, is automatically 'non-contradictory', but information which is contained in the EN cannot be contradicted unless this is specifically permitted by the EN. However, this non-contradictory complementary information may not be incorporated into a National Annex, as would be the most logical and convenient, but must be published as a separate document which the NA may reference. This adds to the plethora of associated documents to which the designer needs access.

It remains to be seen, once EN 1991-1-4 begins to be implemented, what criteria will be used by each relevant authority to judge compliance to these principles. Previous experience suggests that the EN rules will be enforced rigorously in the UK, to the exclusion of best practice or common sense, whereas other Member States may take a more pragmatic approach.

1.5. Design assisted by tests and measurements

With the approval of the appropriate authority, the designer is allowed to obtain load and response information from:

Clause 1.5(1)
Clause 1.5(2)

- wind tunnel tests
- proven and/or properly validated numerical methods
- appropriate full-scale data.

The first two options require *'appropriate models of the structure and of the natural wind'*. The 'appropriate authority' in the UK for buildings is the Building Regulations and, in some circumstances (e.g. during construction), the Health and Safety Executive.

These provisions mirror similar provisions in national standards but, unlike the EN, these standards (such as BS 6399-2[1]) give the necessary criteria to judge compliance. As these clauses are not principles, it would be appropriate for the National Annex to set suitable criteria, if not for the ban on alternative application rules. The only caveat on using experimental data of these three kinds is *'approval of the appropriate Authority'*.

To avoid constant enquiries to these bodies, the UK has interpreted the caveat as an exemption to the general ban on alternative application rules and has inserted the basic criteria to be met into the UK NA. These criteria are the same as in BS 6399-2 for buildings, but now also apply to bridges and other structures. Further guidance is given in the background document.[xx]

UK NA 2.2

1.6. Definitions

In addition to the specific definitions in EN 1991-1-4, the definitions given in ISO 2394,[4] ISO 3898,[5] ISO 8930[6] and ISO 8402[7] also apply. As this is a very large list, it is unreasonable to comment on any but the specifically relevant definitions here. The designer should refer to

EN 1990, clause 1.5 for a basic list of definitions appropriate for this EN, in particular the '*common terms*' and the '*terms relating to actions*'.

1.6.1. Fundamental basic wind velocity

This is the parameter defining the geographical variation of strong winds in a standard exposure, the basic starting-point for estimating the design wind velocity at a site. It is defined as:

> '*the 10 minute mean wind velocity with an annual risk of being exceeded of 0.02, irrespective of direction, at a height of 10 m above flat open country terrain and accounting for altitude effects (if required)*'.

This is the existing standard for most of the Member States, but differs from the current (hourly mean in BS 6399-2[1]) and previous (3 s gust speed in CP3-V-2) standards for the UK. This adopted datum corresponds to the national meteorological datum for most continental European states. Values are given in the National Annex.

1.6.2. Basic wind velocity

This parameter is the fundamental basic wind velocity with the optional directional and seasonal factors applied. Some Member States may choose to ignore directional and seasonal effects by setting the corresponding factors, c_{dir} and c_{season}, to unity. While it is appropriate to include directional effects at the ultimate limit, considering all possible directions, the seasonal effects are only relevant to temporary structures or to serviceability assessments.

1.6.3. Mean wind velocity

This parameter is the 10-minute mean wind velocity at a specified height above ground appropriate for the exposure of the site under consideration. It is equivalent to the site wind speed V_s in BS 6399-2,[1] except that V_s is an hourly mean. It will always include the effects of terrain roughness and orography (hills, escarpments, etc.), and will also include directional and seasonal effects if the NA gives values for c_{dir} and c_{season}.

1.6.4. Pressure coefficient

The pressure coefficients give the effect of the form of the structure on distribution of pressure over the external and internal surfaces. The external pressure coefficients depend only on the external shape of the structure, while the internal pressure coefficients depend on how the external distribution of pressure leaks into the structure.

1.6.5. Force coefficient

The force coefficients give the overall loads on the whole structure, or on specified elements of the structure. In effect, they represent the integration of the surface pressure distribution. Force coefficients may be specified in wind axes, i.e. along-wind (drag) and across-wind (lift) forces, or in body axes, i.e. in fixed directions aligned with the orthogonal structural axes.

1.6.6. Background response factor

This is the factor B that controls the size effect factor, c_s, which allows for the lack of full correlation of fluctuations in the external pressure over the surface of the structure. The background response factor B is not directly used in the EN, so that its inclusion in the list of definitions is quixotic. A definition of size effect factor, c_s, would have been more appropriate as this has an equivalent in BS 6399-2[1] as the size effect factor C_a.

1.6.7. Resonance response factor

This is the factor R that controls the dynamic response factor, c_d, which allows for dynamic response in the fundamental mode of the structure. Generally, this will be the fundamental mode of the whole structure, but it may be appropriate for a major component that responds independently of the whole structure, e.g. a tall chimney stack on top of a building. Again, a definition of dynamic response factor, c_d, would have been more appropriate as this has an equivalent in BS 6399-2[1] as given by $1 + C_r$, from the dynamic augmentation factor C_r.

1.7. Symbols

The symbol notation is based on the definitions in ISO 3898:1999[5] to try to ensure consistency across the Eurocode suite. Three conventions should be noted:

Clause 1.7(1)

- The EN uses the continental convention of a comma to represent a decimal point, so that '1,001' represents 'one, decimal, zero, zero, one' and not 'one thousand and one'.
- The continental convention for multiplication is used: the 'raised dot', e.g. $a \cdot b$, instead of $a \times b$. This is **always** used to avoid confusion with functional dependence.
- Brackets ( ) generally denote a functional dependence on another variable parameter: $v_m(z)$ indicates that the value of v_m depends on the value of z. Where there is more than one functional parameter within the brackets, they are separated by commas, as in $S_L(z_e, n_1)$. However, brackets are sometimes also used to contain an expression group in complex equations. There would be less scope for confusion if the convention of 'curly' brackets had been used exclusively for functional dependence, e.g. $v_m\{z\}$.

As these conventions are so different from UK practice, it is very important that the user takes great care in interpreting equations. The key is to understand the way that the multiplication symbol and brackets are used together. Hence $a \cdot (b - c)$ means a is multiplied by the result of b minus c; whereas $a(b - c)$ means the value of a is a function of b minus c.

Important warning

A full list of the symbols for the parameters used in EN 1991-1-4 is given in clause 1.7(2). The list includes a set of commonly-used subscript indices. Symbols that include subscripts not in this list are listed individually, but there is some apparent duplication, e.g. 'crit' is the standard index for critical, but v_{crit} is specifically defined as the '*critical wind velocity of vortex shedding*'.

Clause 1.7(2)

Most symbols are readily distinguishable from each other, but the Latin symbol v for velocity is quite easily confused with the Greek symbol ν used, variously, for solidity ratio, expected frequency, Poisson ratio and kinematic viscosity. In most cases, the context of the symbol's use will provide the distinction; however, there is one Expression (7.15) in which both v and ν are used together.

The use of w, rather than the obvious choice p, as the symbol for pressure occurs because p is reserved by ISO 3898:1999[5] for probability. In EN 1991-1-4 the symbol p is used to represent '*the annual probability of exceedence*', i.e. the probability that a value is exceeded in any one year. This is not the standard convention in statistics, which is to use $1 - P(x)$. $P(x)$ conventionally represents the probability that x is **not** exceeded, and is called the cumulative distribution function, while $p(x)$ conventionally represents its derivative, the density function dP/dx. To avoid any confusion between p, P and $1 - P$, BS 6399-2 used the symbol Q for probability of exceedence, i.e. $Q(x) = 1 - P(x)$.

CHAPTER 2

Design situations

This chapter is concerned with the various types of design situation that must be addressed in the design of the structure. The material in this chapter is covered by references in *Section 2*, to the design situations defined by EN 1990, clause 3.2:

Clause 2(1)P

- Persistent situations
- Transient situations
- Accidental situations
- Seismic situations

These encompass '*all situations that are reasonably foreseeable or will occur during the execution or use of the structure*'. For example, it is not 'reasonably foreseeable' that a seismic event will occur at the same time as a severe wind storm, so that seismic actions are not relevant to EN 1991-1-4. But it is quite possible that an accidental situation, e.g. vehicle collision, may occur during a wind storm. However, EN 1991-1-4 defines an additional relevant situation:

- Fatigue.

2.1. Persistent situations

These refer to the conditions of normal use which, in this context, refer to the wind actions on the structure in a storm of the characteristic risk. Sustained high winds will occur for only a few hours and the peak wind load relevant to static structures for only about a second, nevertheless these are classed as '*persistent*'.

As structures are required to withstand these persistent loads without distress, it has been the convention to design structures, particularly buildings, under the assumption that they will remain intact, i.e. that the external envelope of a building remains closed. In the UK, the specific need to assess the effect of elective openings, e.g. opening a large door, was introduced only recently as a serviceability limit-state case, but was an implied requirement since the replacement of permissible-stress codes by partial-factor limit-state structural codes. EN 1991-1-4 specifically addresses elective openings only in the context of wind in combination with an accidental load, so it is therefore necessary to ensure that the design loads for the completed enclosed structure are not exceeded during construction or in service.

Persistent wind loads are applied to the structure with the appropriate partial factor γ_f. They may occur in combination with snow and imposed loads, as specified by the combination factors in EN 1990. Fatigue due to the effects of wind should also be considered for susceptible structures.

Clause 2(2)
Clause 2(5)

2.2. Transient situations

Clause 2(3)

These refer to temporary conditions of the structure during construction, assembly or repair. Appropriate values of wind loads need to be determined to ensure that the structure is not overloaded while these transient processes take place. EN 1991-1-6 *Actions during Execution* may seem, by its title, to encompass this situation, but deals principally with the loads that may be imposed on the structure by the construction process itself, e.g. by transporting construction materials around the structure.

The construction process is sufficiently well defined that all relevant situations are easily foreseeable. These will include, for example, unclad frames, unpropped walls and temporary dominant openings. The designer should make full use of directional and seasonal allowances to minimize conservatism in wind loads during the construction phase before allowing an increase in the design risk. Provided that the safety of the workforce and the general public is properly maintained through an active safety management process, the appropriate authority (e.g. the Health and Safety Executive in the UK) may permit a higher annual risk during construction. (In which case, it would be prudent to insure against the increased risk of economic losses due to possible wind-induced failure while the structure remains incomplete.)

Clause 2(2)

These transient wind loads are also applied to the structure with the appropriate partial factor γ_f. They may also occur in combination with snow and imposed loads, as specified by the combination factors in EN 1990. Low-cycle fatigue may need to be considered in the case of slender cantilevered members or other susceptible elements of a partially completed structure. 'Low-cycle fatigue' is the term given to fatigue caused by strain-hardening the structural material through relatively few (hundreds) cycles of loads that are a high proportion (e.g. 80%) of the design load. This may happen to the fixings of cladding or an exposed cantilever member in a wind storm sustained over several hours.

2.3. Accidental situations

These refer to exceptional conditions that are rare or occur for a short time. EN 1990 defines these as '*usually of short duration but of significant magnitude, that is unlikely to occur on a given structure during the design working life*'. For example:

- Debris impacting onto glazing may cause a dominant opening on the windward face of a building, and increase uplift loads on the roof. This implies that an upstream building has already failed in the storm, or that other potential missiles exist immediately upwind.
- A wall may rely on propping by the roof for its stability, but may remain unpropped for several days after the roof is destroyed by fire.

In both of these cases, the risk is lower than the characteristic risk of the design wind velocity because it includes other factors, also of low risk, such that the combined risk of all factors is the appropriate design value. The combined risk of several simultaneous actions is the product of the individual risks when these individual risks are all statistically independent and uncorrelated. These accidental wind loads are applied to the structure with a reduced partial factor, generally $\gamma_f = 1$. While they might occur in combination with snow and imposed loads, it is likely that the National Annex will give combination factors of zero. Note that wind storms and snow are not actually statistically independent and uncorrelated since both are meteorological events occurring in winter.

Clause 2(4)

Clause 2(4) of the EN states: '*Where in design windows and doors are assumed to be shut in storm conditions, the effect of these being open should be treated as an accidental design situation.*' At first sight, this clause could be taken to be sufficient consideration of elective dominant openings, but this is not the case. Clause 7.2.9.1(P), later, sets the principles for internal pressures and it will be seen that these include '*every combination of possible openings*'. As the user of the building has control over whether or not a door is opened, and the door will be opened many times in the working life of the building, so the open situation cannot reasonably be considered an accidental situation. What clause 2(4) means to say is that elective dominant openings should be considered to be open **when assessing accidental loads in combination with wind**. However, while a dominant opening in a building will increase wind loads on the roof and other walls, it reduces the wind load on the wall containing the opening. So even this interpretation should not be considered sufficient, and both open and closed conditions should be considered for the accidental situation.

Important warning

In assessing an accidental situation in combination with wind, the combination factor on the wind loads is only $\psi_{1,1} = 0.2$, so that this is not sufficient to account for the typical service case in wind which must still be considered separately. The assessment of wind loads in service remains a separate important task. Best practice in the UK is given by BRE Digest 436[8] – which is to apply the BS 6399-2[1] probability factor $S_p = 0.85$ in combination with a unity partial factor $\gamma_f = 1$ for serviceability. This will typically provide the critical design case for uplift on long-span roofs when large doors are open on the windward face.

Accordingly, the requirement of clause 2(4) is likely to lead to unsafe designs if wrongly interpreted as being sufficient on its own. The prudent designer will consider the accidental load in combination with both the open and the closed cases, and will also consider both these cases in service conditions. Note that it is the owner/user that becomes responsible for ensuring that elective openings are closed in wind storms exceeding the service condition.

Important warning

2.4. Fatigue

Fatigue effects due to wind actions must be assessed for structures that are susceptible. EN 1991-1-4 gives no guidance on such susceptibility. This is the responsibility of the relevant structural Eurocode.

Clause 2(5)

CHAPTER 3

Modelling of wind actions

This chapter is concerned with the way that wind actions are represented by the Eurocode. The material in this chapter is covered in *Section 3*, in the following clauses:

- Nature   *Clause 3.1*
- Representation of wind actions   *Clause 3.2*
- Classification of wind actions   *Clause 3.3*
- Characteristic values   *Clause 3.4*
- Models   *Clause 3.5*

3.1. Nature

In this context 'modelling' does not mean the construction of physical or numerical scaled models, as required for wind tunnel testing or computational fluid dynamics (CFD) calculations permitted by clause 1.5. It means the 'conceptual model' – the simplified methodology used in the EN to represent the complex processes of wind actions.

Wind actions fluctuate in time and in position over the external surfaces of a structure in a complex and apparently random manner. Stiff structures will respond directly to these actions, allowing the use of the static design model, whereby the maximum strain is directly proportional to the maximum load. Dynamic structures selectively amplify their response to wind actions through resonances at the natural frequencies of the structure. In some circumstances, the motion of the structure is sufficient to increase the wind actions and such structures are called aeroelastic.

The EN uses mathematical models to represent these static, dynamic and aeroelastic action effects as follows: *Clause 3.1(1)*

- *External pressures*: normal pressures over the external surfaces of enclosed structures and the internal surfaces of open structures caused directly by the wind actions.
- *Internal pressures*: normal pressures over the internal surfaces of enclosed structures caused indirectly by the external pressures acting through porosity of the external surface.
- *Normal forces*: forces normal to the surface of the structure cause by the action of the normal pressures.
- *Tangential forces*: forces acting tangentially to the surface caused by friction which may be significant when large areas of structures are swept by the wind.

Pressure is actually a scalar quantity that acts equally in all directions. The term 'normal pressure' is used to imply that these pressures, applied to the surface of a structure, cause forces which act normal to the surface, as opposed to the tangential forces that are caused by friction.

3.2. Representation of wind actions

The EN represents the actual distributions of these pressures and forces by a simplified set of values that give structural loads equivalent to the extreme wind actions, i.e. the most onerous *'characteristic'* loads. The simplification in the model inevitably involves a degree of conservatism to ensure that the most onerous loading case is included. For this reason, design assisted by testing and measurement, as permitted by clause 1.5, often results in lower design loads and a more efficient structure.

3.3. Classification of wind actions

Clause 3.3(1)

These 'characteristic' loads should be applied as *'variable fixed actions'* as specified by EN 1990; that is, as *'an action for which the variation in magnitude with time is neither negligible nor monotonic'*, except where specified. The specified exception is for elective dominant openings (see section 2.3 above) which are treated as *'accidental actions'*. However, in both cases the actions are represented as *'an upper value with an intended probability of not being exceeded'*.

3.4. Characteristic values

Clause 3.4(1)

The wind actions are *'characteristic values'*, i.e. values with a characteristic annual risk of being exceeded of 0.02 in each and every year that the structure remains in service. This level of risk is alternatively described by the mean recurrence interval or 'return period' given by the reciprocal of the annual risk, in this case 50 years.

The use of the term 'return period' is now discouraged because it is often misinterpreted to imply that recurrence is periodic whereas, in reality, recurrence is well described by the Binomial Distribution. If we define $P(r, n)$ as the probability that a wind velocity will be exceeded r times in n years, then:

$$P(r,n) = \frac{n!}{r!(n-r)!} p^r (1-p)^{n-r}$$

Now with the characteristic annual risk of $p = 0.02$, the probability of exceedences in a period of $n = 50$ years is:

No exceedences, $r = 0$: $\quad P(r,n) = 0.98^{50} = 0.364$

One exceedence, $r = 1$: $\quad P(r,n) = \dfrac{50}{1} \cdot 0.02^1 \cdot 0.98^{49} = 0.372$

Two exceedences, $r = 2$: $\quad P(r,n) = \dfrac{50 \cdot 49}{2 \cdot 1} \cdot 0.02^2 \cdot 0.98^{48} = 0.186$

Three exceedences, $r = 3$: $\quad P(r,n) = \dfrac{50 \cdot 49 \cdot 48}{3 \cdot 2 \cdot 1} \cdot 0.02^3 \cdot 0.98^{47} = 0.061$

Four exceedences, $r = 4$: $\quad P(r,n) = \dfrac{50 \cdot 49 \cdot 48 \cdot 47}{4 \cdot 3 \cdot 2 \cdot 1} \cdot 0.02^4 \cdot 0.98^{46} = 0.0145$

The probability of *at least one* exceedence is found from the value for no exceedences, $1 - 0.364 = 0.636$. Hence the chance that the characteristic value is exceeded at least once in the return period is approximately 64%, i.e. almost twice as likely as not. The mean recurrence interval is maintained because years in which there are no exceedences are balanced by years when there are multiple exceedences, so that the average rate of all exceedences is one per return period.

3.5. Models

Clause 3.5(1)

EN 1991-1-4 models the effect of wind on the structure, depending on its size, shape and dynamic properties. This is essentially a *dynamic* model in which the accelerations of the structure are significant. However, the EN simplifies this model into quasi-steady pressures and forces by the use of a dynamic factor c_d that allows static-based design where this is appropriate – which is exactly the same model as in BS 6399-2,[1] where the dynamic augmentation factor C_r is equivalent to $c_d - 1$.

The EN also models *aeroelastic* response in those cases where the motion of the structure modifies the aerodynamic forces. Susceptible structures are cables, masts, chimneys and bridges. Design of potentially aeroelastic structures essentially involves eliminating the possibility of aeroelastic response.

Clause 3.5(2)

CHAPTER 4

Wind velocity and velocity pressure

This chapter is concerned with deriving design values of wind speed and the corresponding velocity pressure for any geographical location and site exposure. The material in this chapter is covered in *Section 4*, in the following clauses:

4.1. Basis for calculation

The peak factor model is used as the basis for calculation in EN 1991-1-4. The principle of this model is that the maximum quasi-steady gust loading or the dynamic response of a structure may be described by a mean, steady part, added to a turbulent, unsteady part. The proportion of the turbulent part is expressed by a peak factor, g, which, for gust loads, depends on the size of the gust and, for dynamics, on the characteristics of the structure. Hence, for the maximum gust speed, $\hat{v}(z)$:

Clause 4.1(1)

$$\hat{v}(z) = v_m(z) + g(t) \cdot \sigma_v(z) = v_m(z) \cdot [1 + g(t) \cdot I_v(z)] \qquad (D4.1)$$

where $v_m(z)$ is the mean wind velocity at height z, $g(t)$ is the peak factor for duration t, $\sigma_v(z)$ is the root-mean-square of the turbulence, so that the term $I_v(z) = \sigma_v(z)/v_m(z)$ is the turbulence intensity. Dynamic response of the structure is accounted for by applying a resonant response factor to the turbulent component.

This model is the basis of calculation of most codes, worldwide, and its development and promotion to this dominant position is attributed to Davenport.[9] When more sophisticated extreme-value models are used to derive the codified data, the resulting values are often cast into this format for consistency, as with the Cook–Mayne methodology[10,11] which is the basis of many of the pressure coefficients in EN 1991-4, the current UK standard BS 6399-2,[1] and other current standards worldwide.

Owing to the squared relationship between wind velocity and pressure, the linear relationship for maximum wind velocity corresponds to a quadratic expression for maximum wind loads of the form:

$$\hat{F}_w = \bar{F}_w \cdot [1 + g(t) \cdot I_v(z)]^2 = \bar{F}_w \cdot \left[1 + 2 \cdot g(t) \cdot I_v(z) + g^2(t) \cdot I_v^2(z)\right] \qquad (D4.2)$$

Being non-linear, equation (D4.2) is inconvenient to deal with when calculating dynamic response, so it is usually linearized to the form:

$$\hat{F}_w \approx \bar{F}_w \cdot [1 + 2 \cdot g(t) \cdot I_v(z)] \qquad (D4.3)$$

by discarding the final squared term in the expanded equation (D4.2). This is a reasonable thing to do when the turbulence intensity $I_v(z)$ is small, i.e. at the top of tall buildings where $I_v(z) \sim 0.1$, but the missing $g^2(t) \cdot I_v^2(z)$ term becomes significant close to the ground in urban areas where $I_v(z) \sim 0.3$.

UK NA 2.17

Important warning

EN 1991-4 adopts the linearized form of equation (D4.3) in the recommended procedures for calculating peak quasi-steady loads and dynamic response, but allows national choice in the case of the peak loads. The UK National Annex uses the full relationship of equation (D4.2) for peak loads to maintain the required factor of safety for low-rise buildings in urban areas. Other National Annexes may adopt the linear form but may also introduce compensating effects (e.g. increasing the value of the peak factor $g(t)$, setting $h_{dis} = 0$ (see sections 4.3.2 and 4.3.5 below) or the choice of turbulence factor k_I (see section 4.4 below), so it is important that the user follows whatever rules are defined in the relevant NA in full.

4.2. Basic values

4.2.1. Fundamental value of basic wind velocity

Clause 4.2(1)P

The fundamental value of basic wind velocity $v_{b,0}$ is defined as the 10-minute mean wind velocity with a 0.02 annual risk of being exceeded, irrespective of direction and season, at 10 m above ground level in terrain Category II.

Category II terrain is defined as open country with low vegetation and isolated obstacles with separations of at least 20 obstacle heights. This corresponds approximately to the World Meteorological Organization's datum exposure for anemometers. EN 1991-1-4 assigns a value of aerodynamic roughness $z_0 = 0.05$ m to Category II terrain, whereas the British Standard BS 6399-2[1] assigns the value $z_0 = 0.03$ m to the same terrain. This difference between datum roughness definitions is not significant provided all adjustments for the other terrain categories are made using the same datum.

The 10-minute averaging period is the meteorological standard for much of continental Europe, but some individual countries use one hour, including the UK and Germany. Both these countries have adopted a factor of 1.06 to adjust the measured one-hour average data to the 10-minute period, based on empirical calibrations.

UK NA 2.4

Clause 4.2(1)P, Note 1

The values of basic wind velocity $v_{b,0}$ are given for each Member State in the corresponding National Annex. The UK NA gives values in a map (Fig. NA.1) that has been adjusted to sea level, defining these as 'map' values, $v_{b,map}$, and introduces an altitude factor c_{alt} to adjust these values to the required base:

$$v_{b,0} = v_{b,map} \cdot c_{alt} \qquad (D4.4)$$

The map in the UK NA is very similar to that in BS 6399-2[1] except that the source data record has been increased from 11 years to 30 years and the original hourly-mean values have been factored up by 1.06 to represent 10-minute mean values.

The UK map values are not directly comparable with the values for the adjacent countries, particularly France, because of the differing choices on application rules made in the respective National Annexes. In particular, France chooses to include altitude and roughness change effects within the design map, whereas the UK extracts these effects from the map and applies them in the application rules. As a consequence, the French map values are higher than those for the corresponding UK coast, i.e. the map value for Calais is significantly higher than the map value for Dover. But when the respective application rules are applied, there is a good match in the resulting design values.

While it may seem perverse to remove roughness fetch effects from the UK map and put them back within the applications rules, this is not actually the case. About a third of the

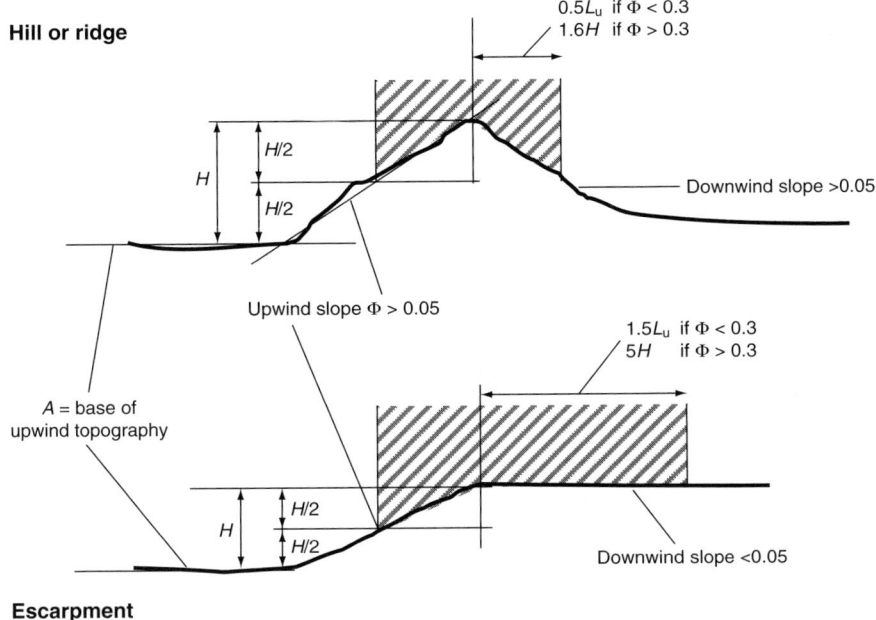

Fig. D4.1. UK rules for significant orography

change in wind speed from 'sea' terrain to 'country' terrain occurs within the first 1 km from the coast, another third between 1 km and 10 km and the final third between 10 km and 100 km. As the UK is an island, the distance from the upwind coast depends on the wind direction, whereas many other Member States have only a single, or no, coastline. Accordingly, the high values along the Atlantic coast of France should decrease more quickly inland than is actually shown on the French map, but this discrepancy is conservative.

The UK altitude factor is given by: *UK NA 2.5*

$$c_{alt} = 1 + 0.01 \cdot A \qquad \text{for } z \leq 10\,\text{m} \qquad\qquad (D4.5a)$$

$$c_{alt} = 1 + 0.01 \cdot A \cdot (10/z)^{0.2} \quad \text{for } z > 10\,\text{m} \qquad (D4.5b)$$

where A is the altitude of the site above mean sea level when orography is not significant, or the altitude of the upwind base of significant orography, and $z = z_s$, as defined in Fig. 6.1 of the EN, or the height above ground of the part above ground in Fig. 7.4 of the EN.

The UK altitude factor is the same empirical factor used in BS 6399-2[1] for buildings, except that the value now decreases with height above ground to allow for taller structures, such as communications masts. Introducing this factor considerably reduces the number of sites where the complex orography rules need to be applied, as shown here in Fig. D4.1, i.e. only in the upper half of hills or close to the crest of an escarpment. The altitude A is taken at the site when orography is not significant and to the base of the hill or escarpment when it is significant – the same rule as BS 6399-2.[1]

4.2.2. Basic wind velocity

The basic wind velocity v_b is the fundamental value $v_{b,0}$ after application of the directional and seasonal, c_{dir} and c_{season}. The recommended values of both these factors is 1.0, but there is the choice in the EN to introduce National Values. The UK NA invokes this choice, giving tables of directional and seasonal factors, unchanged from BS 6399-2.[1] While these UK factors would be expected to be applicable along the Atlantic coasts of France, Belgium and the Netherlands, they may not be used for these locations because the NA for the relevant Member State must be applied. *Clause 4.2(2)P*
UK NA 2.6
UK NA 2.7

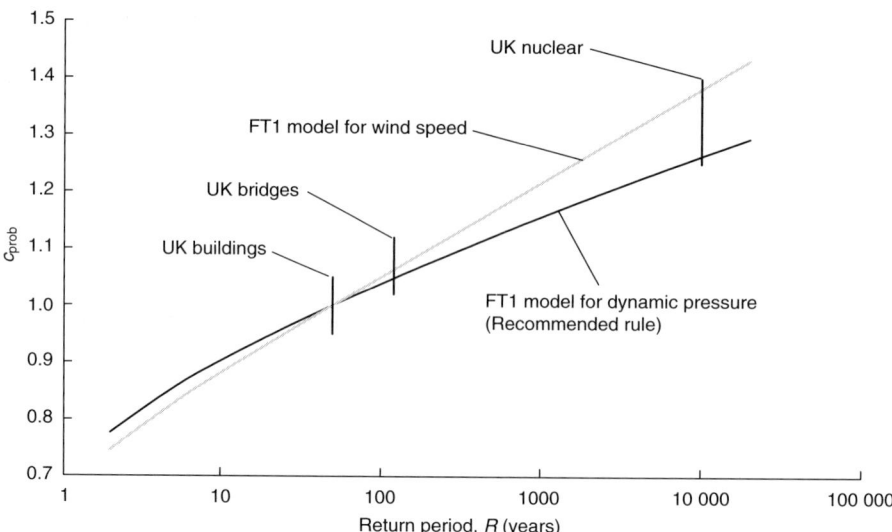

Fig. D4.2. Probability factor models

Clause 4.2(2)P,
Note 2
UK NA 2.6

The UK NA gives values of the directional factor c_{dir} for 30° increments of direction and permits interpolation for intermediate angles. These values are the same as in BS 6399-2.[1] The wind loads on any individual structure are assessed for 'orthogonal cases' because the EN provides pressure coefficients only for these cases (see Chapter 5, later), whereas BS 6399-2 provides pressure coefficients for 30° increments of direction. Accordingly, the appropriate direction factor for each orthogonal case is the highest value found ±45° from the normal wind direction for that case.

Clause 4.2(2)P,
Note 3
UK NA 2.7
Clause 4.2(3)

The UK NA gives values of the seasonal factor c_{season} for any possible one-, two- and four-month period during the year, as well as the six-month summer and winter periods. These values are the same as in BS 6399-2.[1] The main use for the seasonal factor is in assessing wind loads on temporary structures and on structures during construction. Clause 4.2(3) specifically prohibits using the seasonal factor for '*transportable structures which may be used at any time of year*'. The important caveat is '*may be used at any time of year*', implying a lack of control over the period of exposure. However, if the use of a transportable structure is temporary, e.g. a tower-crane assembled and used on site for a specified period, the seasonal factor may be used.

Clause 4.2(2)P,
Note 4
UK NA 2.8

A note to the relevant clause allows the use of a probability factor c_{prob} to adjust the annual risk of being exceeded in expression (4.2). The UK NA invokes the recommended values of $K = 0.2$ and $n = 0.5$, which corresponds to a Fisher–Tippett Type 1 (FT1) distribution of dynamic pressure, so that the probability factor c_{prob} for the UK is identical to S_p in BS 6399-2.[1] Other Member States may choose to adopt the FT1 distribution for wind speed, in which case $K = 0.1$ and $n = 1$. Figure D4.2 shows these two models for the probability factor c_{prob} plotted against return period R where $R \approx 1/p$, i.e. a return period of $R = 50$ years represents an annual risk of exceedence of $p = 0.02$. For common structures in the UK such as buildings and bridges, the difference in the models is small, but for nuclear installations the difference is quite large.

UK practice prior to 1995 was to use the FT1 model for wind speed, but both theory and observations showed that the FT1 model for dynamic pressure used throughout most of Europe is the appropriate model.

4.3. Mean wind

4.3.1. Variation with height

Clause 4.3.1(1)

The mean wind velocity $v_m(z)$ is principally a function of height, through the effects of terrain roughness and orography. It is obtained by multiplying the basic wind velocity v_b by the roughness factor $c_r(z)$ and the orography factor $c_o(z)$. The statement that $c_o(z) = 1$ 'unless *otherwise specified in 4.3.3*' is misleading because it implies this is the typical case. However clause 4.3.3 specifies that orography with slope greater than 3° (5%, or 1:20) should be taken into account, and this will more often than not be the case.

Clause 4.3.1(1)

The UK National Annex adopts the 5% slope threshold, but limits the need to assess $c_o(z)$ to sites close to the crest of hills or escarpments, i.e. sites within the shaded zones shown in Fig. NA2 (reproduced as Fig. D4.1, above).

UK NA 2.5

4.3.2. Terrain roughness
The roughness factor $c_r(z)$ accounts for the effect of the rough ground surface on the vertical profile of wind velocity. Boundary layer theory shows the appropriate profile of mean wind velocity is of the form:

Clause 4.3.2(1)

$$v_m(z) = 2.5 \cdot u_* \cdot \left[\begin{array}{l} \ln\left(\dfrac{z - h_{dis}}{z_o}\right) + 5.75 \cdot \left(\dfrac{z - h_{dis}}{z_g}\right) - 1.875 \cdot \left(\dfrac{z - h_{dis}}{z_g}\right)^2 \\ -\dfrac{4}{3} \cdot \left(\dfrac{z - h_{dis}}{z_g}\right)^3 + \dfrac{1}{4} \cdot \left(\dfrac{z - h_{dis}}{z_g}\right)^4 \end{array} \right] \qquad (D4.6)$$

where u_* is a friction velocity and z_g is the geostrophic height of the boundary layer. The geostrophic height is large in strong winds caused by temperate depressions, $z_g \approx 2000\,\text{m}$, so that only the first logarithmic term is significant close to the ground. The parameter h_{dis} is the optional displacement height parameter, described later in section 4.3.5.

Note that equation D4.6 does not apply to winds caused by other weather mechanisms, such as thunderstorm downburst, katabatic winds, tornadoes or hurricanes. Where necessary, the NA will include any special rules needed for Member States subject to other weather mechanisms. The information given by EN 1991-1-4 is not directly applicable to other countries where the principal strong wind mechanism is not temperate depressions.

Important warning

The EN simplifies this profile to the first logarithmic term, so that $c_r(z)$ becomes:

$$c_r(z) = k_r \cdot \ln\left(\frac{z - h_{dis}}{z_o}\right) \qquad (D4.7)$$

Expression (4.4) of the EN, where $k_r = 2.5 \cdot u_*/v_b$ is a terrain factor, defined by Expression (4.5). The EN also limits the minimum height above ground to z_{min}, depending on the ground roughness, and the maximum height to $z_{max} = 200\,\text{m}$. The recommended values of z_o and z_{min} are tabulated for each terrain category (Table 4.1 of the EN).

Note that equation (D4.7) differs from Expression (4.4) by the inclusion of the displacement height h_{dis} which is defined in the EN as an NV. (See section 4.3.5 below for the definition of h_{dis}.) The EN does not include h_{dis} in any expression because the recommended method does not use it. This is just one of the many inconveniences in the EN which a cynical observer might think were deliberately introduced to frustrate legitimate changes in the NA. As the UK NA chooses to implement h_{dis}, all equations in this Guide include this parameter where relevant.

Figure D4.3 compares the full model of equation (D4.6), used in BS 6399-2[1], with the simplified EN model equation (D4.7) and the 'power law' model commonly used outside Europe for the datum terrain roughness, $z_o = 0.05\,\text{m}$. The EN 'logarithmic law' model is good close to the ground, but becomes increasingly unconservative at levels above ~100 m. The 'power law' model used in some earlier codes of practice remains good at high levels.

A caveat in the EN states that Expression (4.4) is valid '*when the upstream distance with uniform terrain roughness is sufficient to stabilize the profile sufficiently*'. The recommended value for the width of the sector of uniform terrain roughness, defined in Fig. 4.1 of the EN, is ±15° either side of the wind direction, but the choice of the 'upstream distance' is left to the NA.

Clause 4.3.2(2)

The actual behaviour after a change in surface roughness is illustrated in Fig. D4.4(a) for an onshore wind, i.e. a change from sea to land. Wind crossing the sea has a long distance, or

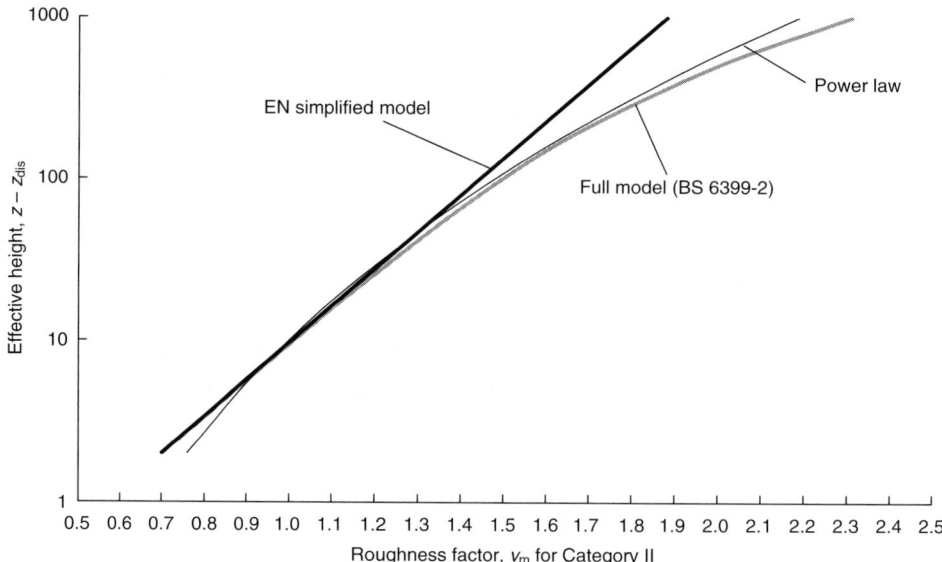

Fig. D4.3. Models for the roughness factor $v_m(z)$

'fetch', over which it establishes an equilibrium profile, where the turbulent stresses balance the surface drag. After crossing the coastline, the increase in surface drag causes the wind velocity near the ground to slow down until a new equilibrium is reached. This effect works gradually upwards through the boundary layer, so that the wind at higher levels does not start to slow down until some distance downwind of the coast.

Design codes often simplify this effect by assuming no change occurs for a certain distance downwind, followed by a sudden change to new equilibrium conditions, as illustrated in Fig. D4.4(b). This simplification is used in the recommended method of EN 1991-1-4, as well as the 'Standard Method' of BS 6399-2.[1] If we first consider the sea-to-land transition, this simplification is very good for most of continental Europe far from the sea, where the effects are small. It is also quite adequate for those countries with a single coastline where errors in the model can be mitigated in the specification of basic wind velocity. However, it is different for the UK which, being a small island, has a coastline at a different distance

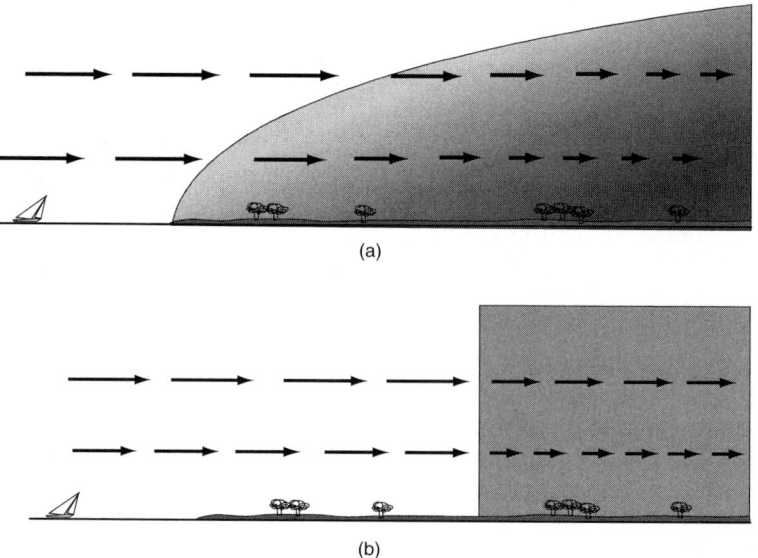

Fig. D4.4. Response of atmospheric boundary layer to a change of ground roughness: (a) actual behaviour; (b) recommended model in EN 1991-1-4

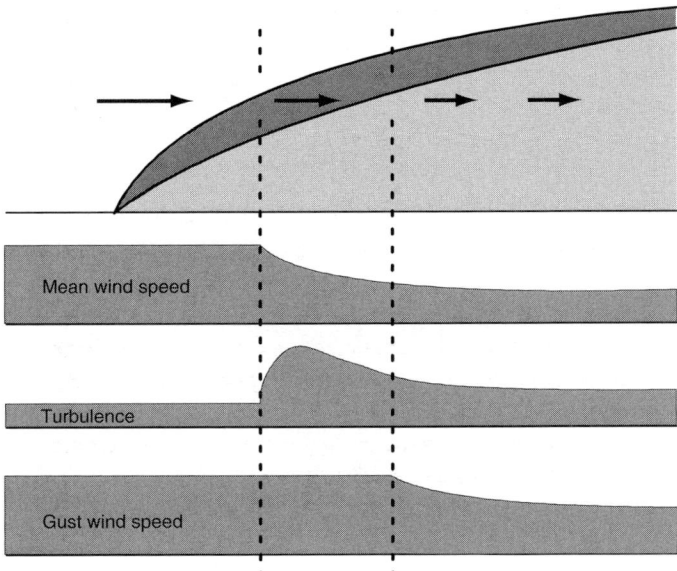

Fig. D4.5. Transitional effects through the roughness boundary

in every direction: sites near west-facing coasts will experience westerly winds stronger than expected for the datum roughness, while sites by east-facing coasts will experience stronger easterly winds.

The same effect occurs after a transition from rural, or 'open country' roughness, to urban, or 'town' roughness, but here the higher urban turbulence intensity can have a significant effect on the dynamic response of tall buildings near the urban boundary. Figure D4.5 enlarges the transition zone near the change of roughness to show that it has a significant thickness. In order for the kinetic energy of the mean wind velocity to be absorbed it must first be converted to turbulence before it is transferred by the turbulent Reynolds stresses to the ground surface, and this takes some time. Figure D4.5 shows that mean wind velocity starts to decrease at the front edge of the transition zone, while gust velocity does not decrease until after the rear edge of the zone. Through the zone, the turbulence intensity first increases to a high value before settling down to the new equilibrium value. From this it follows that buildings close to an urban boundary will experience gust values similar to the previous open country, but higher levels of turbulence and therefore enhanced dynamic response.

The design process for rectangular-plan buildings is often simplified to four orthogonal cases, or less if there is symmetry in the structure. Each orthogonal case implies a 90°-wide range of wind direction, i.e. three times the recommended width. Accordingly, clauses 4.3.2(3) and (4) require the user to adopt the smallest roughness found within any 30°-wide sector within the range considered. *Clause 4.3.2(3)* *Clause 4.3.2(4)* *UK NA 2.12*

While most countries are expected to define a moderate 'upstream distance' of a few kilometres, the UK NA defines a distance of 100 km and provides a method that accounts for all intermediate values. Implementation is simplified by combining some of the terrain categories: *UK NA 2.11*

- Terrain Category 0 is referred to as 'Sea' terrain.
- Terrain Categories I and II are considered together and referred to as 'Country' terrain.
- Terrain Categories III and IV are considered together and referred to as 'Town' terrain.

The roughness factor effects then depend on the upwind 'distance to sea' and 'distance in town' and are implemented by charts:

- Figure NA3 of the EN gives the roughness factor $c_r(z)$ directly for all sites, from effective height above ground, $z - h_{dis}$ and 'distance upwind to shore line'.

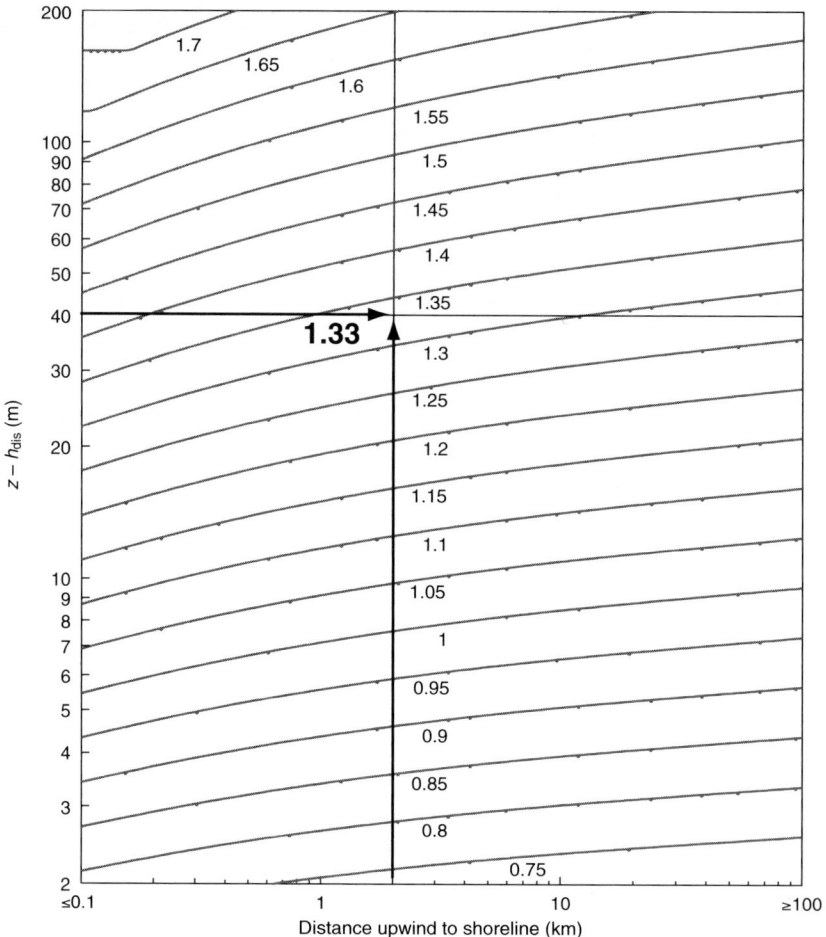

Fig. D4.6. Roughness factor look-up chart in UK National Annex

- Figure NA4 of the EN gives a correction factor for urban sites, i.e. Categories III and IV, from effective height above ground, $z - h_{dis}$ and 'distance inside town terrain'.

The UK implementation reduces the effort of determining the roughness factor $c_r(z)$ while reducing unnecessary conservatism by implementing a more complex roughness model.

The simplicity of the UK NA implementation is illustrated by Fig. D4.6 which reproduces Fig. NA3. Values of $c_r(z)$ are read by interpolating between the contours on this chart. For this example, the value of $c_r(z)$ for $z - h_{dis} = 40$ m and a distance 2 km from the coastline is 1.33. However, implementing the UK NA is simplified further by the software tool illustrated in Fig. D4.7, which may be obtained free of charge by downloading from www.rwdi-anemos. com.

Currently, the UK design rules of BS 6399-2[1] include large inland lakes in the category 'Sea' when the site is close to the lake – '*inland areas of water extending at least 1 km in the wind direction when closer than 1 km upwind of the site*'. The EN includes lakes in Category I, so that the UK NAD rules now incorporate them into 'Country' terrain, giving correspondingly lower design wind speeds. Buildings immediately adjacent to a lake shore will have no permanent obstructions upwind for winds blowing off the lake ($h_{dis} = 0$), which limits the size of this change. Nevertheless, the EN rules are non-conservative in this context, although the deficiency does not entirely erode the standard partial load factor, γ_f. Accordingly, the UK NAD requires that all lakes extending more than 1 km in the wind direction should still be treated as 'Sea' when the site is closer than 1 km upwind.

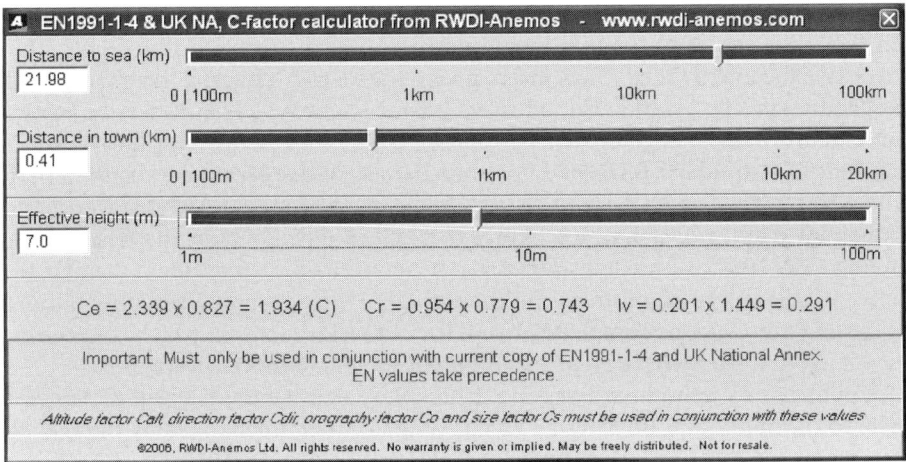

Fig. D4.7. Free calculator tool for EN 1991-1-4 wind speed and turbulence factors

4.3.3. Terrain orography

Clause 4.3.3(1) requires orography, i.e. mountains, hills, escarpments and cliffs, to be assessed when they increase wind velocities by more than 5%. This is not a very helpful instruction because it requires an assessment to be made in order to see whether an assessment needs to be made. The recommended procedure to do this is given in Annex A.3. Although clause 4.3.3(2) specifies that orography may be neglected when the '*average slope of the upwind terrain*' is less than $3°$ (corresponding to a slope of 0.05, or 1 in 20), this will not apply for many sites.

Clause 4.3.3(1)
Clause 4.3.3(2)

However, the statement that '*the upwind terrain may be considered up to a distance of 10 times the height of the isolated orographic feature*' may cause the user some confusion. The '*distance of 10 times the height of the isolated orographic feature*' should be interpreted as the minimum distance that the site can be from the crest of the feature before the feature can be discounted. At the minimum hill slope of 0.05, or 1 in 20, orography must be considered if the site lies in the upper half of the hill. But, for a hill slope of 0.1 (1 in 10) or steeper, orography must be considered if the site lies anywhere on the hill.

The UK National Annex reduces the need to implement the complex rules for assessing orography by implementing the optional altitude factor, c_{alt}, described earlier in section 4.2.1. The site need only be within the zones shown in Fig. D4.1 to require full assessment, although full assessment is permitted anywhere near the feature and may give less onerous wind velocity.

UK NA 2.13

4.3.4. Large and considerably higher neighbouring structures

It is well known that low-rise buildings can be adversely affected by strong winds which are brought down to ground level by the adjacent tall buildings, but this effect has not been previously included in national or international design codes. This deficiency was addressed by a draft model code published by the European Convention for Constructional Steelwork (ECCS).[12] EN 1991-1-4 includes the simpler of two methods recommended by this model code.

Clause 4.3.4(1)
UK NA 2.14

The method is given in Annex A.4 (see section 9.2 later) and should be implemented when the structure being designed is close to another that is at least twice its height.

4.3.5. Closely spaced buildings and obstacles

When buildings or other permanent obstacles are packed closely together, they provide mutual shelter in a zone extending from the ground to about the average level of roof-tops. The presence of this sheltered zone causes the wind profile to rise upwards, giving an effective ground level just below the average level of rooftops at a height called the displacement

Clause 4.3.5(1)

height h_{dis}. This displacement reduces the wind velocity at any given height above ground to less than the value expected for that height without obstructions. In this Guide, the height above this displacement height has been given the name 'effective height', but has not been assigned a specific symbol. So, effective height is always given as $z - h_{dis}$ to reinforce the requirement to assess h_{dis}.

The current UK standard BS 6399-2[1] permits the use of displacement height in permanent woodland. The critical issue here is the meaning of 'permanent'. Many woodland areas are commercial plantations that are likely to be clear felled within the design lifetime of the building. EN 1991-1-4 avoids this issue by confining displacement height to urban areas only. While most urban areas increase in size, there remains the possibility that an extensive urban area may be completely razed for redevelopment, in which case the exposure of buildings around the periphery of this area will be made more severe.

UK NA 2.15

The recommended method for assessing the displacement height is given in Annex A. This is identical to the method in BS 6399-2,[1] so has been adopted unchanged by the UK NA. The method assesses the average effect of all obstacles over a distance of some 100 m upwind of the site. It therefore discounts the direct shelter given by any one individual building which would disappear if the sheltering building were demolished. Nevertheless, the user needs to consider whether these obstacles are permanent. It is generally assumed that urbanization is a continuously increasing process, so that buildings on the urban boundary eventually become surrounded by other buildings.

However, in the event of large-scale demolition and redevelopment, a building may become more exposed to the wind for a significant period. The recommended method limits the reduced wind loads to around 70% of the value without obstructions, i.e. a reduction of 30%. Following demolition of upwind buildings, the increase in loading would significantly erode, but not eliminate, the partial load factor for wind, $\gamma_f = 1.4$.

Use of the displacement height h_{dis} reduces the design wind speed by reducing the effective height above ground, so setting $h_{dis} = 0$ is always a safe option. Some National Annexes may set $h_{dis} = 0$ to provide a conservative error in order to compensate for the non-conservative error in using the linearized gust factor model, described earlier in section 4.1.

4.4. Wind turbulence

Clause 4.4(1)

Turbulence intensity $I_v(z)$ at any height is the standard deviation (rms) of the wind velocity divided by the mean value, so $I_v(z) = \sigma_v(z)/v_m(z)$. The standard deviation of the turbulence close to the ground is relatively constant with height, so that the EN adopts the simplified model whereby the intensity decreases with height in inverse proportion to the increase in mean wind velocity between the heights of z_{min} and z_{max} and a constant value below z_{min}. The model also assumes that the standard deviation of the turbulence is not affected by acceleration of the mean wind velocity over orography, i.e. the flow over orography complies with the rapid distortion theory of Jackson and Hunt,[13] so that the turbulence intensity decreases in inverse proportion to the increase in mean wind velocity over the orography.

The recommended Expression (4.7) in the EN is given here as:

$$I_v(z) = \frac{k_I}{c_o(z) \cdot \ln\left(\dfrac{z - h_{dis}}{z_o}\right)} \qquad \text{for } z_{min} \leq z - h_{dis} \leq z_{max} \qquad (D4.8a)$$

$$I_v(z) = I_v(z_{min}) \qquad \text{for } z < z_{min} \qquad (D4.8b)$$

where the effective height $z - h_{dis}$ has been used. The recommended value of the turbulence factor $k_I = 1.0$ but this value is subject to National Choice.

Equations (D4.8a) and (D4.8b) complex equations and the implication is that it provides an accurate estimate of the turbulence. However, setting the recommended value $k_I = 1.0$ and $c_o(z) = 1.0$ and $h_{dis} = 0$ for flat open terrain, we obtain $I_v(z) = 1/\ln(z/z_o)$. Since $c_r(z) = k_r \cdot \ln(z/z_o)$, we find that $c_r(z) \cdot I_v(z) = k_r$, so that the rms turbulence is predicted to be constant value for all heights.

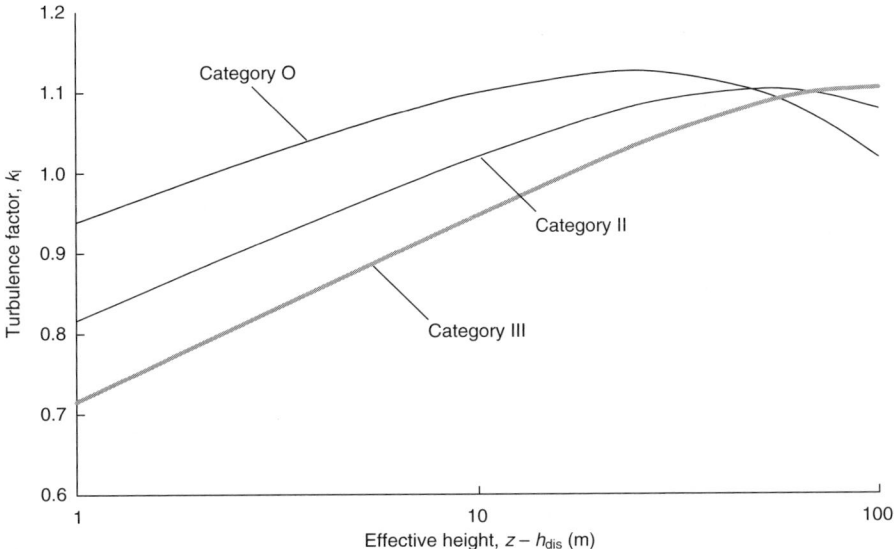

Fig. D4.8. Equilibrium values of turbulence factor, k_I

In reality the rms turbulence is constant near the ground, but reduces significantly with height. This model of equations (D4.8a) and (D4.8b) gives quite a poor representation of equilibrium turbulence conditions with height and entirely fails to predict the enhanced turbulence levels just after a change to rougher terrain. Better equilibrium values of k_I are given in Fig. D4.8, showing that k_I is also a function of height.

Figure D4.9 compares the vertical profile of turbulence intensity with effective height above ground of the full model adopted by the UK NA to the simplified model of EN 1991-1-4, $k_I = 1.0$. Below 10 m above ground, the simplified model is conservative, but between 10 m and 100 m it is non-conservative. Above 100 m it becomes conservative again. Accordingly, the simplified model alone is not sufficient to compensate for the non-conservative error from using the linearized gust factor model (see section 4.1 above). In order to avoid all issues of 'compensating errors' the UK NA adopts the full models in all cases, offsetting any complexity of these models by the use of design charts.

As the EN defines k_I as an NV, the UK NA presents its choice of value directly as the value for $k_I/\ln[(z - h_{dis})/z_o]$ in Expression (4.7), so eliminating the need for further calculation when $c_o = 1$.

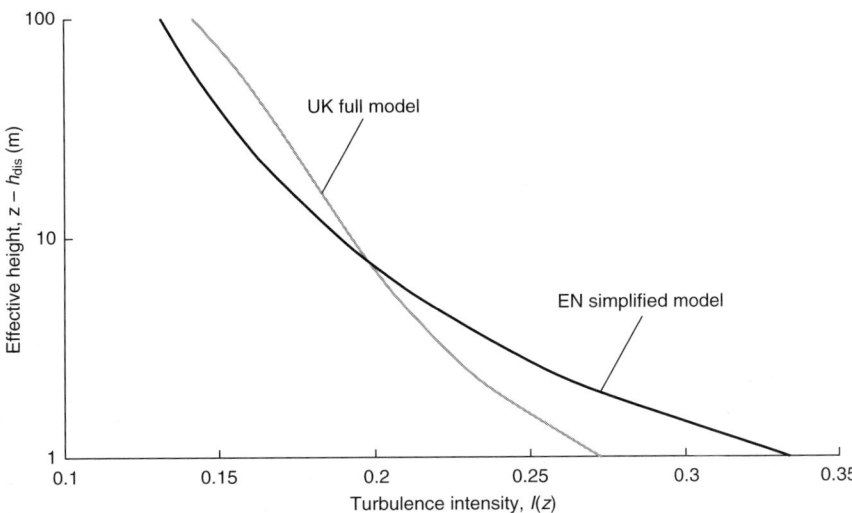

Fig. D4.9. Turbulence intensity profile models for Category II terrain

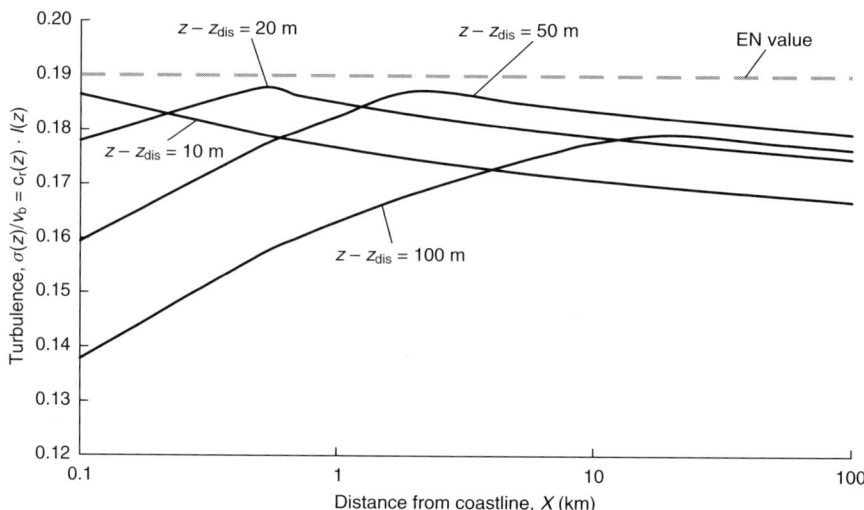

Fig. D4.10. Development of turbulence downwind of coastline at a number of heights in Category II

UK NA 2.16

Immediately after a smooth-to-rough transition at the coastline or urban boundary, the turbulence is initially higher than the equilibrium values, as shown in Fig. D4.5. The UK National Annex includes the effects of roughness transitions on the turbulence intensity in the combined factor $k_I/\ln[(z - h_{dis})/z_o]$ using the same graphical format as used for the roughness factor $c_r(z)$. The UK NA provides two charts:

- Fig. NA5 giving the factor $k_I/\ln[(z - h_{dis})/z_o]$ directly for all sites, from effective height above ground, $z - h_{dis}$ and *'distance upwind to shore line'*.
- Fig. NA6 giving a correction factor for urban sites, i.e. Categories III and IV, from effective height above ground, $z - h_{dis}$ and *'distance inside town terrain'*.

In flat terrain, where $c_o(z) = 1$, the values from the charts give the turbulence intensity $I_v(z)$ directly. Otherwise, where topography is significant, the values must be divided by $c_o(z)$. The UK implementation reduces the effort of determining the turbulence intensity while using a more complex roughness model that accounts for the enhanced turbulence immediately after a change to rougher terrain.

Figure D4.10 illustrates how the turbulence develops after a change of roughness from 'Sea' to 'Land' at various heights. The rms turbulence $\sigma_V(z)$ is plotted as a ratio of the base wind speed v_b, obtained from the product of the turbulence intensity $I_v(z)$ and the roughness factor $c_r(z)$. It is seen that close to the ground, the rms turbulence falls from an initial high level towards the equilibrium value. Higher above ground, a longer fetch of land is required to reach the maximum rms turbulence. Figure D4.10 gives values to the diagrammatic representation of the transitional effects in Fig. D4.5. The corresponding value in the EN is a constant for each terrain category, as noted above. The resulting constant EN value for Category II is $\sigma_V(z)/v_b = 0.19$, which Fig. D4.10 shows is a conservative simplification.

4.5. Peak velocity pressure

Clause 4.5(1)

Unlike previous codes of practice, like BS 6399-2,[1] the EN does not specify gust velocities, which would need to be converted to dynamic pressures for equivalent static design of structures, but moves directly to the required gust dynamic pressure. The EN calls this the 'peak velocity pressure' $q_p(z)$ to avoid any possible confusion over the use of the term 'dynamic' which is reserved in the EN for describing the dynamic response of the structure. Values of the peak velocity pressure will be required for the majority of low- and medium-rise structures and for elements of dynamic structures.

The EN splits derivation of the peak velocity pressure into the two steps:

- The basic velocity pressure: $q_b = \frac{1}{2} \cdot \rho \cdot v_b^2$
- The exposure factor: $c_e(z) = q_p(z)/q_b$

The basic velocity pressure is simply the basic mean wind speed expressed as a dynamic pressure, so that the exposure factor is the factor required to adjust this for site exposure.

Earlier factors, $c_r(z)$ and $c_o(z)$, are factors on *velocity*, but $c_e(z)$ is a factor on *pressure*. The only distinction between these is the subscript, and this is likely to cause some confusion. To avoid the possibility of confusion, the UK standard, BS 6399-2, uses S for factors on velocity and C for factors on pressure, and it would have been sensible for the EN to adopt a similar approach. *Important warning*

The recommended rule given by Expression (4.8) is the linearized version of the peak factor model (equation (D4.1)), described earlier in section 4.1. The value of peak factor for the shortest measured gust ($t \approx 1$ s) is generally accepted to be $g(t) = 3.5$ based on a datum averaging period of one hour. EN 1991-1-4 adopts this value without taking into consideration that the extreme 10-minute mean wind speed, used as reference, is around 6% higher than the hourly-mean value.

While it may have been common practice within continental Europe to use $g(t) = 3.5$ with the 10-minute averaging period, the UK, in common with most of the rest of the world, has always used this value with a one-hour averaging period. Adoption of the recommended rule, Expression (4.8), would overestimate gust wind loads by around 12% close to the ground. On the other hand, the linearized version of the peak factor model would underestimate gust wind loads close to the ground, i.e. a compensating error. Rather than accept these partially compensating errors, the UK NA replaces Expression (4.8) by the full model and the appropriate value of gust factor: *UK NA 2.17*

$$q_p(z) = [1 + 3.0 \cdot I_v(z)]^2 \cdot \frac{1}{2} \cdot \rho \cdot v_m^2(z) = c_e(z) \cdot q_b \qquad (D4.9)$$

(expression NA3).

The constant of 7 used in the linearized Expression (4.8) equates to $2g(t)$, so that the value $g(t) = 3.5$ used in the full model is the peak factor value generally taken to apply with an hourly-mean reference wind velocity.

Figure D4.11 compares the values of exposure factor $c_e(z)$ for Category II roughness from:

- the UK NA full model (equation (D4.9))
- the EN simplified linear model with a peak factor of $g(t) = 3.5$ (Expression (4.8), where the constant $7 = 2g(t)$)
- the EN simplified linear model with a peak factor of $g(t) = 3.0$ (the UK NA value).

This comparison shows that using the hourly-mean value of $g(t) = 3.5$ in the linear model, instead of the 10-minute mean value of $g(t) = 3.0$, adds about 5% to the value of

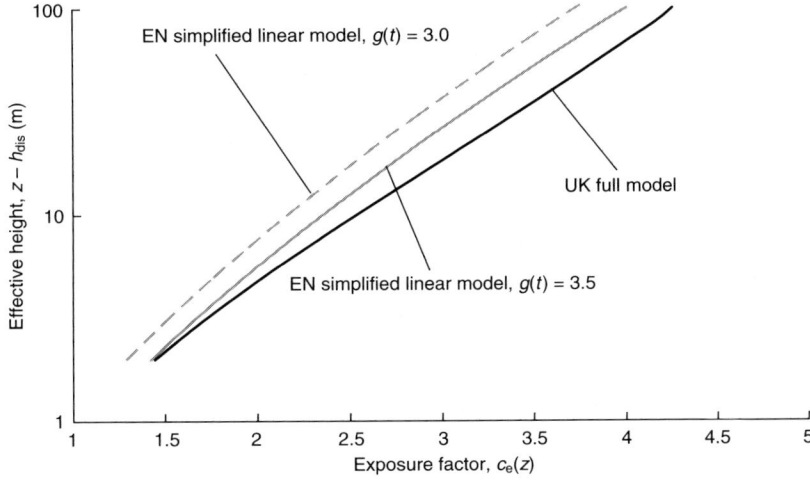

Fig. D4.11. EN and UK NA models for exposure factor $c_e(z)$ for Category II (100 km from coastline)

$c_e(z)$. However, this is not quite enough to compensate for the non-conservative linearization.

The UK NA also provides design charts to determine the exposure factor directly for flat terrain, $c_{e,flat}(z)$, in the same consistent format used for the roughness and turbulence intensity factors. These apply when $c_o(z) = 1$, i.e. for flat terrain. The UK NA gives an expression to correct these values for the effects of orography as:

$$q_p(z) = q_b \cdot c_{e,flat}(z) \cdot (c_o(z) + 0.6)/1.6 \qquad (D4.10)$$

(Expression (NA4)) which is generally conservative for heights above ground up to 50 m. For heights above 50 m, Expression (NA3) should be used.

The user of the UK NA has the choice of determining the peak velocity pressure from the mean wind velocity and turbulence intensity, using two/four charts and equation (D4.8), or directly from the basic velocity pressure, using one/two charts and equation (D4.9), with equation (D4.10) if needed. The first choice is more appropriate if the structure is significantly dynamic, because the user will require the mean wind velocity and turbulence intensity to calculate the dynamic response. The second choice is more appropriate for static structures.

UK NA 2.18   The recommended value of air density is $\rho = 1.25\,\mathrm{kg/m^3}$, which is relatively high and relates to very low temperatures at low altitude. The UK NA adopts a value of $\rho = 1.226\,\mathrm{kg/m^3}$, which is more appropriate for strong winds blowing off the Atlantic Ocean.

Summary of wind velocity parameters compared with BS 6399-2[1]

Parameter name	EN 1991-1-4 General	EN 1991-1-4 UK National Annex	BS 6399-2 Parameter name	BS 6399-2
Uncorrected fundamental basic velocity	Allows altitude effects	$v_{b,map}$	Basic wind speed	V_b
Altitude factor		c_{alt}	Altitude factor	S_a
Fundamental basic velocity	$v_{b,0}$	$v_{b,0} = v_{b,map} \cdot c_{alt}$	n/a	n/a
Direction factor		c_{dir}	Direction factor	S_d
Seasonal factor		c_{season}	Seasonal factor	S_s
Probability factor		c_{prob}	Probability factor	S_p
Basic wind velocity	$v_b = v_{b,0} \cdot c_{dir} \cdot c_{season} \cdot c_{prob}$		Site wind speed	$V_s = V_b \cdot S_a \cdot S_d \cdot S_s \cdot S_p$
Displacement height	Allowed	h_{dis}	Displacement height	H_d
Height above ground	z	$z - h_{dis}$	Effective height	$H_e = H - H_d$
Minimum height	z_{min}		Minimum effective height	$H_e \geq 0.4 \cdot H$
Orography factor	$c_o(z)$	$c_o(z - h_{dis})$	Topography increment	S_h
Roughness factor	$c_r(z)$	$c_r(z - h_{dis})$	Fetch and turbulence factors	S_c, T_c
Mean wind velocity	$v_m(z)$	$v_m(z - h_{dis})$	Mean wind speed	$V_o = V_s \cdot S_c \cdot T_c$
Turbulence factor	k_I			
Turbulence intensity	$I_v(z)$	$I_v(z - h_{dis})$	Factors S_t and T_t	$S_t \cdot T_t$
Air density	$\rho = 1.25\,\mathrm{kg/m^3}$	$\rho = 1.226\,\mathrm{kg/m^3}$	Air density	$\rho = 1.226\,\mathrm{kg/m^3}$
Basic velocity pressure	$q_b = \frac{1}{2} \cdot \rho \cdot v_b^2$			
Exposure factor	$c_e(z)$	$c_e(z - h_{dis},\,\mathrm{flat})$	(Terrain and building factor)^2	S_b^2 (Standard)
		$c_e(z - h_{dis})$		S_b^2 (Directional)
Peak velocity pressure	$q_p(z)$	$q_p(z - h_{dis})$	Dynamic pressure	$q = \frac{1}{2} \cdot \rho \cdot (V_s \cdot S_b)^2$

CHAPTER 5

Wind actions

This chapter is concerned with the actions on structures caused by the wind, as covered by *Section 5*, in the following clauses:

- General   *Clause 5.1*
- Wind pressures on surfaces   *Clause 5.2*
- Wind forces   *Clause 5.3*

5.1. General

Like all wind codes worldwide, EN 1991-1-4 requires that the actions of wind on a structure should take account of the actions on both external and internal surfaces by means of the external and internal surface pressures. The net pressure across any external surface will be the difference between these external and internal pressures. The net pressure across any internal partition will be the difference between internal pressures on either side of the partition.

Clause 5.1(1)P

- External pressures are generated directly by the flow of wind around the structure.
- Internal pressures are generated by the balance of flow through openings and porosities in the envelope of the structure, driven by the distribution of external pressures over its surface.

Clearly, internal pressures do not exist when the structure does not enclose a volume of air; for example, lattice towers and boundary walls have no 'inside'. The distinction between 'internal' and 'external' becomes blurred when structures have large openings and the EN adopts the same conventions as BS 6399-2.[1]

- Where the structure encloses a space and has one or more faces formed by permanent walls, but some faces are entirely open, it is deemed to have an 'inside' and an 'outside', and requires both internal and external pressures to be determined. Examples include sports grandstands and aircraft hangars.
- Where the structure does not enclose a space, or has no faces formed by permanent walls, net pressures are directly defined across each surface. Examples include free-standing walls and canopy roofs over petrol stations.

5.2. Wind pressures on surfaces

The wind pressures acting on the external surfaces, w_e are obtained by multiplying an external pressure coefficient c_{pe}, obtained from *Section 7*, by the peak velocity pressure $q_p(z_e)$ at the reference height z_e, using Expression (5.1):

Clause 5.2(1)

$$w_e = q_p(z_e - h_{dis}) \cdot c_{pe} \qquad (D5.1)$$

Clause 5.2(2)

Similarly, the wind pressures acting on the internal surfaces are obtained by multiplying an internal pressure coefficient c_{pi}, obtained from *Section 7*, by the peak velocity pressure $q_p(z_e)$ at the reference height z_e, using Expression (5.2):

$$w_i = q_p(z_e - h_{dis}) \cdot c_{pi} \tag{D5.2}$$

The required reference height is always given in the relevant table of *Section 7*. When the National Annex allows use of the displacement height h_{dis}, the effective value of the reference height is reduced to $z_e - h_{dis}$, hence equation (D5.1) and equation (D5.2) replace z_e with $z_e - h_{dis}$ in the expressions given by the EN.

Clause 5.2(3)

The net pressure acting on a surface is simply the difference between the pressures on either side:

$$w = w_e - w_i \tag{D5.3}$$

5.3. Wind forces

Clause 5.3(1)

The wind forces acting on the whole structure, or a component of a structure, are determined in either of two ways:

- indirectly, by summing the components of surface pressures and friction stresses over the structure, or
- directly, by means of force coefficients appropriate to the whole structure or by vector summation of forces on elements of the structure.

When a force coefficient relates to the whole structure there will be an associated reference area, A_{ref}. This is usually the area logically associated with the force, for example the area of the windward face of a rectangular-plan building in the case of overall loads in the direction of the wind (drag). An exception to the rule applies to lattice trusses. For three- and four-boom lattice trusses, the reference area is the solid area of elements in a *single reference face* of these trusses. The value of force coefficient always matches the definition of reference area, so that c_f for a four-boom truss will be larger than c_f for a three-boom truss with the same solidity, because the four-boom truss has an additional face attracting wind loads.

Clause 5.3(2)

The general case for overall wind load by vector summation of forces on a structure is given by Expression (5.4)

$$F_w = c_s c_d \cdot \sum_{elements} [c_f \cdot q_p(z_e) \cdot A_{ref}] \tag{D5.4}$$

that is, expressed as the sum of the load on all elements multiplied by the structural factor, $c_s c_d$, defined in *Section 6*, which accounts for the combined action of size effect and structural dynamics. (It will be seen, later, that the EN allows the structural factor $c_s c_d$ to be split into its component parts, the size factor c_s and the dynamic factor c_d.) The specific case of one monolithic structure, with a single force coefficient, Expression (5.3), is given by this general case with only one element.

Clause 5.3(3)

The overall wind force is obtained by summing the components of external and internal pressures, and the friction stresses, given from Expressions (5.5), (5.6) and (5.7) by:

$$F_w = c_s c_d \cdot \underbrace{\sum_{surfaces} (w_e \cdot A_{ref})}_{\text{External forces}} + \underbrace{\sum_{surfaces} (w_i \cdot A_{ref})}_{\text{Internal forces}} + \underbrace{c_{fr} \cdot q_p(z) \cdot A_{fr}}_{\text{Friction forces}} \tag{D5.5}$$

It is important to note that the size and dynamic effects of the structural factor, $c_s c_d$, are confined to the external components, because the EN assumes that the internal pressures and frictional stresses are steady values that are fully correlated over the surfaces. This assumption is good in respect of the internal pressures, which depend on the internal volume, but does not hold for the frictional stresses. Just as the local fluctuations of external normal pressures caused by the smallest gusts do not act simultaneously across a large surface, the friction effects of these smallest gusts do not act simultaneously either.

For overall forces on enclosed buildings, the internal forces would be expected to cancel out. But for net forces across a building face, where internal pressures are important, EN 1991-1-4 gives no allowance for the effect of building volume on internal pressure.

The friction forces only act on surfaces parallel to the wind and are small compared with pressure forces, so they only become significant when the area of surfaces parallel to the wind is very large. Accordingly, clause 5.3(4) allows the effects of friction to be ignored when the total area of surfaces parallel to the wind is less than four times the total area of windward and leeward surfaces. EN 1991-1-4 assumes that the friction loads are fully correlated and gives no allowance for the non-simultaneous action of gusts over the surfaces parallel to the wind. BS 6399-2[1] includes the size effect in the effective wind speed for both normal pressures and frictional stresses. The assumption of full correlation for frictional stresses in the EN is conservative but, as the frictional contribution on long-span buildings is typically ~10% of the overall force, and much less for typical buildings, the effect of this conservatism is usually not large. However, the frictional stresses dominate the drag force for flat-roofed canopy structures, and in this case the conservatism may be significant.

Clause 5.3(4)

The force obtained by summing the pressures on the windward or the leeward face represents the maximum instantaneous load for that face. However, these two maximum values are unlikely to occur simultaneously, due to lack of correlation between the pressure fluctuations on the windward and leeward faces. Accordingly, the maximum total load is less than the sum of the maximum face loads. The note to clause 5.3(5) allows National Choice on whether to allow for this effect generally or, as recommended later in clause 7.2.2(3), to restrict any allowance to walls only. The UK National Annex allows the reduction factor, defined in NA 2.19, to be applied to the horizontal force component from all windward and leeward surfaces, i.e. on both wall and roof surfaces. In the case of a pitched roof with wind nominally normal to the eaves, the front face from the front eaves to the ridge is interpreted as 'windward' and the rear face from the ridge to the rear eaves is interpreted as 'leeward'.

Clause 5.3(4)
UK NA 2.19

Summary of wind action parameters compared with BS 6399-2[1]

	EN 1991-1-4		BS 6399-2	
Parameter name	General	UK National Annex	Parameter name	
Reference height		z_e	Reference height	H_{ref}
Peak velocity pressure	$q_p(z_e)$	$q_p(z_e - h_{dis})$	Dynamic pressure	$q = \frac{1}{2} \cdot \rho \cdot (V_s \cdot S_b)^2$
External pressure coefficient		c_{pe}	External pressure coefficient	C_{pe}
External pressure	$w_e = q_p(z_e) \cdot c_{pe}$	$w_e = q_p(z_e - h_{dis}) \cdot c_{pe}$	External pressure	$p_e = q_e \cdot C_{pe}$
Internal pressure coefficient		c_{pi}	Internal pressure coefficient	C_{pi}
Internal pressure	$w_e = q_p(z_e) \cdot c_{pi}$	$w_e = q_p(z_e - h_{dis}) \cdot c_{pi}$	Internal pressure	$p_e = q_e \cdot C_{pi}$
Net pressure		$w = w_e - w_i$	Net pressure	$p = p_e - p_i$
Size factor	$c_s \cdot c_d$	c_s	Size effect factor	C_a
Dynamic factor		c_d	1 + dynamic augmentation factor	$1 + C_r$
Reference area		A_{ref}	Loaded area	A
External forces		$F_{w,e} = c_s c_d \cdot \sum (w_e \cdot A_{ref})$	External loads	$P_e = \sum (p_e \cdot C_{a,e} \cdot A)$
Internal forces		$F_{w,i} = \sum (w_i \cdot A_{ref})$	Internal loads	$P_i = \sum (p_i \cdot C_{a,i} \cdot A)$
Area parallel to wind		A_{fr}	Area swept by wind	A_s
Friction forces		$F_{fr} = c_{fr} \cdot q_p \cdot A_{fr}$	Frictional forces	$P_f = q_s \cdot C_f \cdot A_s \cdot C_a$
Overall horizontal force		$F_w = 0.85 \cdot F_{w,e} + F_{w,i} + F_{fr}$	Overall horizontal load	$P = 0.85 \cdot \sum (p_e \cdot C_{a,e} \cdot A) \cdot (1 + C_r)$

CHAPTER 6

Structural factor $c_s c_d$

This chapter is concerned with determining the dynamic response of structures in the fundamental mode of vibration, through the structural factor $c_s c_d$ or its components – the size factor c_s and the dynamic factor c_d – when the National Annex permits separation. The material in this chapter is covered by *Section 6*, in the following clauses:

• General	*Clause 6.1*
• Determination of $c_s c_d$	*Clause 6.2*
• Detailed procedure	*Clause 6.3*

6.1. General

The structural factor accounts for the combined effect of:

Clause 6.1(1)

- non-simultaneous action of peak wind pressures over faces of the structure, generally called the 'size effect'; and
- vibration of the structure in its fundamental mode due to the action of turbulence, generally called the 'dynamic response'.

Note that non-simultaneous action between forces on windward and leeward faces is not included in the structural factor, and is accounted for by a simple calibration factor in clause 7.2.2(3), later in the EN.

Non-simultaneous action of gusts over faces of structures tends to reduce the maximum instantaneous pressure averaged over the surface. This has been known since 1884, when Benjamin Baker measured the forces on plates of different sizes exposed to the wind in preparation for designing the Forth rail bridge. Baker discovered that the smaller plates experienced higher wind loads in proportion to their size than the larger plates and correctly attributed the effect to the size of wind gusts relative to the plates.

On the other hand, resonance of flexible structures at their natural frequencies of vibration tends to amplify their response to fluctuating loads, so that the two effects encompassed by the structural factor tend to compensate each other.

The note to clause 6.1(1) allows National Choice as to whether the structural factor can be separated into two parts:

- the size factor, c_s
- the dynamic factor, c_d.

It is sensible to take these two effects together for the overall response of whole structures, since this simplifies the computational task. However, the model used in EN 1991-1-4 to represent the dynamic response component is valid only for the response of building structures in the first cantilever mode, and is not applicable to bridges or for many of the individual elements of building structures which can be regarded as being static.

Accordingly, separation into size factor and dynamic factor allows size effects to be taken into account for bridges and structural components of buildings.

UK NA 2.20 The UK National Annex permits separation of the structural factor into its two parts and it is expected that other NAs will also permit separation. Table NA3 gives values of size factor c_s depending on the sum of the width and height, $b + h$ of the element. This differs from BS 6399-2[1] which uses the diagonal dimension $a = \sqrt{(b^2 + h^2)}$. Figure NA9 gives values of dynamic factor c_d for various typical classes of structure.

For the dynamic response of buildings in the fundamental modes, the UK user may choose to use the combined structural factor $c_s c_d$ of the EN, or to apply the separate parameters, size factor c_s and dynamic factor c_d. There should be no difference between the results of these choices because the methodology is consistent. Other rules are provided for bridges in the UK NA, derived from current UK practice, as described later in Chapter 8. However, for structures outside the scope of the EN and for components of structures, it is better to use the separate parameters. For components considered to be static, $c_d = 1$ and only the size factor c_s should be used. For other dynamic structures outside the current scope of the EN, the user will need to determine an equivalent dynamic factor c_d, consistent with the EN Principles, or to seek external guidance.

6.2. Determination of $c_s c_d$

Clause 6.2(1) Clause 6.2(1) defines a number of cases where the value of $c_s c_d = 1$ may be assumed, avoiding the calculation process entirely. These are small structures, or elements of structures, where the size effect and dynamic effect are both small, but tend to compensate each other. These include:

- buildings less than 15 m high
- building elements with natural frequencies greater than 5 Hz
- framed buildings with structural walls less than 100 m high, where the height is less than four times the in-wind depth ($h/d < 4$)
- circular chimneys less than 60 m high, where the height is less than 6.5 times the diameter ($h/d < 6.5$).

This assumption is generally conservative, so the use of 'may' in clause 6.2.1 permits the detailed procedure of clause 6.3.1 to be used for these cases, in the expectation that this will give $c_s c_d < 1$.

With the exception of bridges, which are considered as a separate case in section 8, the detailed procedure of clause 6.3.1 should be used to determine the structural factor $c_s c_d$
Clause 6.2(1), unless the National Annex requires the structural factor to be split into its component
Note 1 parts, namely size factor c_s and dynamic factor c_d (for example, in the UK). Annex F gives guidance on determining the natural frequencies of the structure (see section 9.6 below).
Clause 6.2(1), Annex D gives values of $c_s c_d$ for the overall response of various common structural types
Note 2 which comply with these rules, which may be used to avoid calculation. These are described as 'envelopes', implying that they are upper bound values and that they will be generally conservative.

6.3. Detailed procedure

6.3.1. Structural factor $c_s c_d$

Clause 6.3.1(1) The calculation procedure assesses the dynamic response of a structure in the along-wind direction as the root-sum-square of a 'background' and a 'resonant' component. The background component represents the quasi-steady (i.e. not amplified) response of the structure to the atmospheric turbulence, while the resonant part represents the dynamic oscillation of the structure at its natural frequencies. This is usually called the Davenport method.[9]

Confusion may arise because the term 'reference height' is used for a number of different purposes. Here, the method for $c_s c_d$ requires some representative values of overall wind

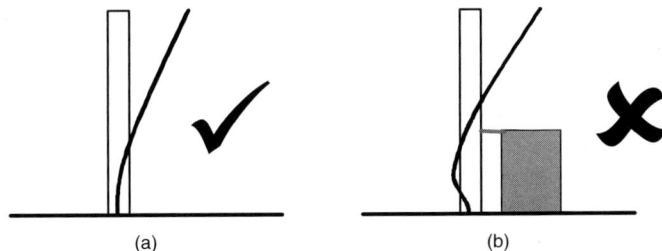

Fig. D6.1. Fundamental mode shapes for chimney stack: (a) cantilever mode; and (b) propped mode

effects on the structure which are calculated at the datum reference height designated by the symbol z_s and defined in Fig. 6.1 of the EN for three categories of structure. **It is very important to note that the value of reference height $z_s = 0.6 \cdot h$ for the *vertical structure like buildings etc* refers only to the wind parameters needed to calculate $c_s c_d$** (and c_{alt} in the UK NAD – see section 4.2.1 above). When it comes to calculating the velocity pressure for use with the pressure coefficients in section 7, a reference height, designated by the symbol z_e, is defined that always corresponds to the top of the structure, or structural part. **Serious underestimation of design wind loads will occur if the definition of z_s in Fig. 6.1 is erroneously carried forward into the calculation of velocity pressure.**

Important warning

Important warning

This reference height z_s is used to derive wind parameters representative of the dynamic response of the structure. In reality, these parameters are integrated over the mode shape (e.g. height) of the structure, so these single-height values are a simplified model. While this is quite adequate in typical situations, there are a number of situations where use of a single reference height may not be appropriate. For example, Fig. D4.10 shows that the rms turbulence, which drives the dynamic response, rises to a peak value at some height above ground, depending on the distance from the coastline (and also the distance into an urban area), so that the reference height defined in Fig. 6.1 of the EN may not be conservative. Similar considerations apply to sites on the crest of hills. In these situations, the user should seek expert advice.

Expression (6.1) gives the combined effect as the structural factor $c_s c_d$, while Expressions (6.2) and (6.3) give the component's size factor c_s and dynamic factor c_d, respectively. When these expressions are used, Note 3 allows values of key parameters to be specified in the National Annex. Alternative procedures are given in Annex B – the well-established Davenport methodology,[9] with some modifications, which was the method used in the ENV; and in Annex C – a newer methodology (from Dyrbye and Hansen[14]) which is claimed to give values within 5% of the other method. One might reasonably ask why it is necessary to give two alternative methodologies when it is stated that they match to within 5% – and the answer is not unconnected to national rivalries within the drafting process.

Clause 6.3.1(1), Note 3

Each National Annex must state which one, or both, of these rival methods should be adopted as normative in the Member State. The UK NA adopts Annex B as normative, with the proviso that the value of k_p determined by the method may not be less than 3.5. Table NA3 giving values of the structural factor c_s and Fig. NA9 giving values of dynamic factor c_d have been compiled using Annex B. The UK NA stipulates that Annex C should not be used in the UK.

UK NA 2.21

The method given by Expression (6.1) is only valid for the along-wind response of a structure in the fundamental mode, and then only when the mode shape has a constant sign. This is expressed as a 'principle'. Accordingly, as shown in Fig. D6.1 above, the response of the chimney stack in the cantilever mode (a) can be assessed by the method as the deflection is exclusively down-wind (constant sign), but the propped mode (b) cannot be assessed as the mode shape includes a region of reversed sign.

Clause 6.3.1(2)P

6.3.2. Serviceability assessments

For the assessment of serviceability, EN 199-1-4 requires that the maximum along-wind acceleration at any given height is considered in addition to the maximum in-wind deflection

Clause 6.3.2(1)

to ensure that the motion experienced by occupants remains within acceptable limits. A method is given in each of the two alternative annexes, namely Annex B and Annex C.

UK NA 2.22 As noted above, each National Annex must state which one, or both, of these rival methods should be adopted as normative. The UK NA adopts Annex B as normative and stipulates that Annex C should not be used.

6.3.3. Wake buffeting

Clause 6.3.3(1) When slender buildings ($h/d > 4$) and chimneys ($h/d > 6.5$) are arranged in groups, the turbulence generated by the wind flowing around upwind structures adds to the gustiness of the wind and increases the dynamic response of downwind structures affected by their wakes; that is, upwind structures cast a turbulent wake onto a downwind structure and the increased dynamic response this causes is called 'wake buffeting'.

Clause 6.3.3(2) Wake buffeting is negligible and need not be considered in the design when the separation between structures is greater than 25 times the cross-wind breadth or when the natural frequency of the downwind structure, subject to wake buffeting, is higher than 1 Hz.

If the limits defined by clause 6.3.3(2) are not exceeded, so that wake buffeting should be considered, the EN recommends wind tunnel tests or specialist advice.

CHAPTER 7

Pressure and force coefficients

This chapter is concerned with defining the pressure and force coefficients that depend only on the external shape of the structure and are independent of location and structural form. The material in this chapter is covered by *Section 7*, in the following clauses:

The scope of this section is supposed to include all structures except bridges, which are treated separately in *Section 8*. However, most other structural forms other than buildings are excluded from the current scope of the EN, so that the scope of *Section 7* is effectively confined to buildings.

There is, however, a great deal of inconsistency in the range of the data provided. A significant range of common building shapes is missing, or inadequately covered, for which reliable data are available in current codes (e.g. BS 6399-2[1]), in published guidance or peer-reviewed journals. Less important information, for which there is little or no provenance, is given equal prominence, for example see section 7.12 'Flags', below. The background document[xx] to the UK NA contains further guidance.

7.1. General

The aerodynamic effects of the wind loads on structures are defined using pressure and force coefficients, as defined earlier in sections 1.6.4 and 1.6.5. Depending on the form of the structure, it will be appropriate to use the following:

Clause 7.1(1)

• *Internal and external pressure coefficients* – for the distribution of normal wind stress over the internal and external surfaces. These are needed when the distribution of loads on surfaces is required, e.g. cladding loads. In some instances, the internal pressure coefficient may be determined from the distribution external pressure coefficient and the permeability of the structure.

- *Net pressure coefficients* – for the net normal wind stress across a surface. These are needed when the structure has no 'inside', so that both sides of a surface are exposed to the wind, e.g. canopies and boundary walls.
- *Friction coefficients* – for tangential wind stresses in the direction of the wind caused by friction, or by the accumulated normal stresses on protrusions on the surface (e.g. ribs or corrugations).
- *Force coefficients* – for overall loads on structures or loads on individual elements, where it is not necessary to define the spatial distribution.

7.1.1. Choice of aerodynamic coefficient

Clause 7.1.1(1)–(4)

This section directs the user to the appropriate section, giving the aerodynamic coefficients for each type of structure. Note 2 explains that the EN gives two types of external pressure coefficient, namely 'overall' and 'local', which are described later in section 7.2.1. Another Note explains that force coefficients represent the overall effect of the normal and tangential wind stresses on a structure, i.e. including friction if this is not specifically excluded.

7.1.2. Asymmetric and counteracting pressures and forces

Clause 7.1.2(1)

The complex patterns of loading are reduced into the simplified form required by codes and represented as zones of constant pressure, so that information on asymmetrical loads tends to be lost. EN 1991-1-4 simplifies the external pressure coefficients for buildings to four orthogonal cases where the coefficient zones and values are symmetric around the notional orthogonal direction. Accordingly, these EN coefficients predict no torsional effects. To make up for this deficiency, special account needs to be taken for structures that are sensitive to asymmetries, e.g. arched or shell structures, or torsion cores.

Clause 7.1.2(2)

   Clause 7.1.2(2) directs the user to specific rules for '*buildings, free-standing canopies and signboards*'. While the procedures for free-standing canopies and signboards give rules for asymmetry, this is not comprehensively covered in the procedures for buildings – so the reference here to 'buildings' is not correct. The Note to this clause gives:

- a recommended procedure for torsion in buildings (Fig. 7.1 of the EN)
- a general, conservative procedure to account for beneficial, or '*counteracting*', loads.

   The EN recommended procedure to account for torsion in buildings is given in Fig. 7.1, reproduced here as Fig. D7.1(a) below. Essentially, the EN rule imposes a linear taper on the windward face pressures to induce a net torsion. Calibration of this rule shows that it underestimates the asymmetric loading caused by wind directions between the orthogonal cases. However, there is National Choice on adoption or replacement of this rule. UK

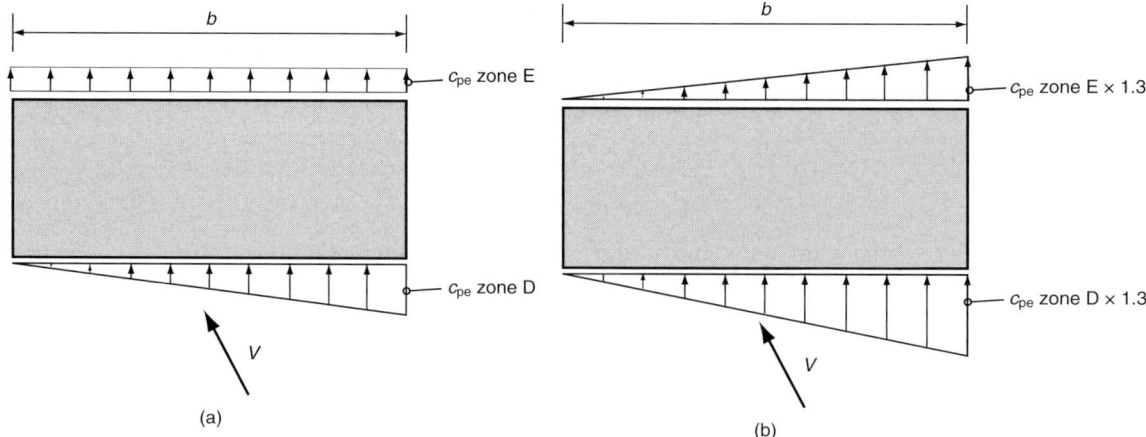

Fig. D7.1. Comparison of recommended and UK asymmetry rules for torsion effects on buildings: (a) EN 1991-1-4 asymmetry rule; and (b) UK NA asymmetry rule

NA addresses this problem by applying a linear taper on both windward face and leeward face pressures as defined in Fig. NA 10 and reproduced here as Fig. D7.1(b). Other NAs may adopt similar or different rules, such as offsetting of the centre of force from the torsional centre by some specified amount.

UK NA 2.23

The National Annex may give other rules to account for asymmetric and counteracting loads.

7.1.3. Effects of ice and snow

The pressure and force coefficients depend critically on the external shape of the structure. This clause reminds the user that build-up of ice and snow may make significant changes to the area or shape of a structure. In particular, fine lattice structures may become so blocked as to be effectively solid. The NA may give specific rules and data. The UK NA gives no further guidance on the subject other than to point the user to EN 1993-3-1 for guidance on lattice towers and masts.

Clause 7.1.3(1)

UK NA 2.24

7.2. Pressure coefficients for buildings
7.2.1. General

This clause states that the external pressure coefficients depend on the size of the loaded area under consideration. In fact, it is the external pressures that depend on the size of the loaded area, through the 'size effect' introduced in section 5.3 above and described in detail in Chapter 6. It is quite possible to define external pressure coefficients that are almost insensitive to size effect when the size factor c_s is separated out from the structural factor $c_s c_d$, as is the case in BS 6399-2.[1]

Clause 7.2.1(1)

On the assumption that size factor c_s will not be separately defined, the EN defines two classes of external pressure coefficient:

- The '*overall coefficients*', which apply to loaded areas of $10\,m^2$ and are intended for determining structural loads, where the recommended application rules for size effect and dynamic response are used in combination, i.e. as $c_s c_d$.
- The '*local coefficients*', which apply to loaded areas of $1\,m^2$ and are intended for determining the loads '*on small elements and fixings*', i.e. on cladding loads.

The local coefficients exist principally because of a reluctance to reduce the very high local coefficients that are in the current codes of some Member States, despite good evidence that they are excessive. They are needed only because the recommended application rules for size and dynamic response, $c_s c_d$, apply to the overall structural loads and the response of the whole structure in the fundamental mode, so cannot cope with small-format cladding elements. The recommended application rules allow interpolation between the '*overall*' and '*local*' values for areas between $1\,m^2$ and $10\,m^2$ on a logarithmic basis.

The UK NA avoids this problem by separating out the size factor c_s and allowing its application with the overall coefficients for all loaded areas larger than $1\,m^2$, exactly as in BS 6399-2.[1] As a consequence, the interpolation rule between $1\,m^2$ and $10\,m^2$ given by Fig. 7.2 of the EN is inappropriate, and may not be used in the UK because this would result in double-counting of the size effect. If the choice were permitted, the UK NA would not adopt the 'local' coefficients at all – but no National Choice is permitted here. Thus, in the UK implementation, the overall coefficients are used with the size effect factor c_s for all loaded areas greater than $1\,m^2$, and the local coefficients are used without the size effect factor for loaded areas of $1\,m^2$ or less.

UK NA 2.25

The external pressure coefficients for buildings are presented for the orthogonal wind directions 0°, 90°, 180° and 270° and should account for the largest values in the range of wind direction ±45° either side of the orthogonal direction. So these coefficients are directly equivalent to the 'Standard method' coefficients of BS 6399-2[1] – indeed the overall coefficients are derived from the same directional source data as BS 6399-2 (see Reference 15) – but the values may differ slightly due to minor differences from redefinition of the zones.

Clause 7.2.1(2)

The first draft of EN 199-1-4 provided detailed directional coefficients, as in the BS 6399-2 Directional method, in an Annex – but these were removed in later drafts to reduce the size and complexity of the EN. However, since both codes are derived from the same source data, the BS 6399-2 Directional values may reasonably be regarded as 'non-contradictory complementary information'.

Clause 7.2.1(3)

The special provision for roof overhangs is the same as in BS 6399-2, in that the pressure on the adjacent wall is taken to apply to the underside (the soffit). There may be other situations where it will be appropriate to adopt this as a more general rule, e.g. for mullions or ribs on a wall. BS 6399-2 defines a threshold size to distinguish a large overhang from a small open-sided building, but the EN does not, leaving this choice to the judgement of the user.

The criterion for sizing the pressure coefficient zones on buildings is consistent across most of the building shapes. The key size parameter e, is defined as $e = b$ or $e = 2 \cdot h$, whichever is the smaller. For example, $e = b$ for tall towers and $e = 2 \cdot h$ for low long-span buildings. This is the same general rule as used in BS 6399-2.

7.2.2. Vertical walls of rectangular plan buildings

Clause 7.2.2(1)

The reference height, z_e, used to determine the velocity pressure for walls, is defined here as the 'upper heights' of the different parts of the walls, which is consistent with current UK practice. (See the 'important warning' in section 6.3.1 above.)

A procedure is given for splitting *windward* walls of tall buildings into a number of different parts. This is a variant of the 'division by parts rule' first introduced in the 1972 revision of the UK wind code (CP3 Chapter V Part 2) as a load-reducing concession, when it was thought that the previous 1970 code was too onerous for overall loads. This rule was immediately seized on by users seeking a way of reducing wind loads on cladding elements. The deficiencies in interpretation of this rule were noticed almost immediately and correctional guidance was given in the 1974 BRE *Wind Loading Handbook*,[16] which was to restrict its application to overall horizontal loads and exclude its use altogether for local 'cladding' loads.

UK NA 2.26

When the UK division by parts rule was first proposed for inclusion in EN 1991-1-4 it was initially rejected by all other Member States on the grounds that it was not universally applicable, but was later accepted after the scope of the rule had been restricted to windward walls only. However, the Note allows the NA to specify other rules for the leeward and side walls, but recommends that the reference height for the leeward and side walls should be taken as the height of the building, $z_e = h$. The UK NA adopts the recommendation of this Note, which confines the 'division by parts' rule to the windward face only.

Clause 7.2.2(2)

Windward and leeward walls of buildings are each taken to be a single zone of constant valued coefficient, while the side walls are divided into vertical strips from the upwind corner. The width of each strip is sized using the parameter e. Account is taken of the height/depth ratio of the building for windward and leeward values. This is identical to the procedure in the Standard method of BS 6399-2.[1] With the option to use different reference heights for the windward face:

- external pressure zones on the windward wall will comprise horizontal strips (constant coefficient, varying velocity pressure)
- external pressure zones on the side walls will comprise vertical strips (varying coefficient, constant velocity pressure), and
- the leeward wall comprises a single pressure zone (constant coefficient and velocity pressure).

Clause 7.2.2(2), Note 2

The values of pressure coefficient c_{pe} for walls are given in Table 7.1 for various proportions of height to depth, h/d, the 'span ratio'. This differs slightly from current UK practice,[1] which only distinguishes between $h/d \geq 1.0$ and $h/d \leq 0.25$, with interpolation permitted between these limits. The EN alternative values for $h/d > 5.0$ assumes these to correspond to tall, slender buildings likely to be dynamic and so falling outside the scope of the UK code.

Clause 7.2.2(3)

The equivalent overall drag coefficient, i.e. force coefficient in the wind direction, is obtained by vector summation of the windward and leeward pressure coefficients. The maximum pressure coefficient on the windward face (zone D) and the minimum (maximum suction) on the leeward face (zone E) do not occur at the same wind direction in the $\pm45°$ range of wind direction, so never act simultaneously and their sum exaggerates the overall drag load.

The UK NA corrects this deficiency by replacing Table 7.1 with Tables NA4a and NA4b. Table NA4a provides the pressure coefficients for the windward (D) and leeward (E) walls and adds the net pressure coefficient for their simultaneous combination, '*net*'. These coefficients depend on the proportions of the building, d/b and h/b. Table NA4b provides the pressure coefficients for the three side wall zones, A, B and C, for the two cases '*Isolated*' and '*Funnelling*'. These coefficients do not depend on the proportions of the building. Here, the term 'funnelling' applies to funnelling of wind between adjacent buildings, and the UK NA gives application rules in the Notes to the tables which comply with current UK practice. Essentially, the '*Funnelling*' values apply on side walls facing another building, when the gap between the two buildings is in the range $>e/4$ and $<e$, where e is the lesser of $e = 2h$ or $e = b$, *and* where both buildings stick up above the general level of the upwind buildings. The '*Isolated*' values correspond to all other cases.

UK NA 2.27

There will inevitably be occasions where structural calculations require the net load to be partitioned between windward and leeward faces, for example in the design of a portal frame. Two critical load cases may be deduced from the values given in Table NA4a:

(1) When the load on the windward face is at the maximum value $c_{pe}(D)$ given in column D, the corresponding simultaneous value for the leeward face cannot be greater than the vector sum $c_{pe}(E) = c_{pe}(D) - c_{pe}(net)$.
(2) When the load on the leeward face is at the maximum value $c_{pe}(E)$ given in column E, the corresponding simultaneous value for the leeward face cannot be greater than the vector sum $c_{pe}(D) = c_{pe}(net) - c_{pe}(E)$.

Thus, for $h/d = 1$ and $h/b = 2$, we have for case 1:

$$c_{pe}(D) = +0.8 \quad \text{and} \quad c_{pe}(net) = 1.1, \text{giving } c_{pe}(E) = +0.8 - 1.1 = -0.3$$

while for case 2 we have:

$$c_{pe}(E) = -0.4 \quad \text{and} \quad c_{pe}(net) = 1.1, \text{giving } c_{pe}(D) = 1.1 - 0.4 = 0.7$$

and the choice of critical design case for any structural element will depend on the value of the internal pressure.

Note 2 to Table 7.1 of the EN (and Note 6 to Table NA4a of the UK NA) allows the force coefficients for tall slender buildings with $h/d > 5$ to be determined using the rules for structural sections in 7.6 *et seq*. This provision provides the only way that the user can account for the reduction in overall drag force and moment caused by rounding the corners of a tall building (see section 7.6 below). When the overall horizontal force in the wind direction is determined by summation of the external pressures, a reduction in overall load caused by lack of correlation between windward and leeward faces is allowed. The Note gives the following recommendations:

- For tall buildings where $h/d \geq 5$, the factor is unity and there is no reduction in load.
- For squat buildings where $h/d \leq 1$, the overall force may be reduced by the factor 0.85.
- In the range $1 < h/d < 5$ the factor may be linearly interpolated.

This is almost equivalent to current UK practice in BS 6399-2,[1] where the factor 0.85 is allowed for all proportions of building. This factor accounts for the lack of correlation, the non-simultaneous action of fluctuating pressures, on front and rear faces, while the size factor c_s accounts for correlation across each individual face. The 0.85 value implies that the pressure on the rear face is close to the mean when the pressure on the front face is at its maximum and vice versa. The rule for tall buildings implies that the fluctuations remain fully correlated due to the action of vortex shedding, but this is a conservative assumption.

7.2.3. Flat roofs

Any roof with a pitch angle less than 5° is defined in the EN as a flat roof. The zones on roofs are divided into windward corner and eaves zones, upwind and downwind main zones, using the size parameter e. The reference height z_e is always to the highest point of the roof, i.e. to the top of any parapet. Values of coefficient are given for sharp eaves, parapets, curved eaves and mansard eaves. In the downwind zones, where the flow has reattached to the roof, the pressure fluctuates around the zero value, and pressure coefficients here are assigned the value ±0.2. The more onerous sign (depending on internal pressure or other structural considerations) is the appropriate value to use. These procedures comply with current UK practice.

7.2.4. Monopitch roofs

Any single pitch roof with a pitch angle of 5° or greater is defined as a monopitch roof. Roof zones are divided into windward corner and eaves/verge zones, upwind and downwind main zones, using the parameter e. The reference height z_e is always to the highest eaves of the roof. Values of coefficient are given for sharp eaves only. These procedures comply with current UK practice, except that BS 6399-2[1] gives additional rules for monopitch roofs with parapets, curved and mansard eaves.

7.2.5. Duopitch roofs

Any double-pitch roof, whether ridged or troughed, with a pitch angle of 5° or greater is defined as a duopitch roof. However, the clause gives values only for pitches of equal slope. Where the pitches are not of equal slope, but are reasonably similar, the pitch of the upwind slope should be used as datum. Roof zones are divided into windward corner eaves/verge and ridge zones, upwind and downwind main zones, using the parameter e. The reference height z_e is always to the highest point of the roof. Values of coefficient are given for sharp eaves only. These procedures comply with current UK practice, except that BS 6399-2[1] gives additional rules for roofs with parapets, curved and mansard eaves and guidance where each pitch can have a different angle.

7.2.6. Hipped roofs

All hipped roofs with a pitch angle of 5° or greater are covered by this clause, even if the pitch angles differ between the main and hip slopes. The datum pitch angle is always the pitch of the windward slope, as indicated by Note 3. Roof zones are divided into windward corner eaves/verge and ridge zones, upwind and downwind main zones, using the parameter e. The reference height z_e is always to the highest point of the roof. Values of coefficient are given for sharp eaves only. These procedures comply with current UK practice, except that BS 6399-2[1] gives additional rules for roofs with parapets, curved and mansard eaves and guidance where each pitch has a different angle.

7.2.7. Multispan roofs

Pressure coefficients on multispan roofs are determined from the pressure coefficients for a single-span roof of the same type and pitch, with an allowance for reduction in values for downwind spans. The rules are essentially identical to current UK practice in BS 6399-2.[1] Figure 7.10 of the EN shows the interpretation of the rules for multispan monopitch (sawtooth) and multispan duopitch roofs. In the latter case, all pitches between the upwind and downwind ridges are treated as being troughed, with a negative pitch angle. While multispan hipped roofs are not specifically addressed, the rules for multispan duopitch roofs may be applied to this case.

7.2.8. Vaulted roofs and domes

The original source, and hence the provenance, of the EN values for vaulted roof and domes is not known. It is not believed to be the only set of data known to have been obtained in a simulated boundary layer, which were measured at a Brazilian university in 1990[15] and only give mean values. Nor does it include later published data, because the values appear to have been copied from pre-existing national codes. Because of these deficiencies, EN 1991-1-4 allows National Choice for these values.

Because barrel-vault roofs are currently a popular form of roof for light-industrial buildings, a study to validate these values was recently undertaken in the UK by the Building Research Establishment. This used exactly the same analysis procedures as had been used to derive the source data for the wall, flat and pitched roof coefficients. This study showed significant differences from the values in Fig. 7.11 of the EN. Accordingly, the UK NA provides new, better values for barrel vaults in Figs NA10 and NA11 using the same presentation format as Fig. 7.11. The study did not address the companion data for domes, so the UK NA retains the recommended values for domes through lack of any reliable alternative.

UK NA 2.28

7.2.9. Internal pressure

The value of the internal pressure of a building is set by the balance of flow through openings into the building driven by the distribution of external pressure. The principle is that internal and external pressures shall be taken to act at the same time and that the worst combination *'shall be considered for every combination of possible openings and leakage paths'*. This requirement is a principle, so acts as a 'catch-all' which includes the case of elective dominant openings, as noted earlier in section 2.3 and discussed in more detail below. It puts the onus on the designer to anticipate the pattern of use to which the building may be put by the eventual owner (see section 1.3) above. However, the 'every combination' requirement is not as onerous at it might appear because there are limits to the range of internal pressures which are determinate. One limit is set by enclosed buildings with small openings, the other by open-sided buildings or buildings with large dominant openings.

Clause 7.2.9(1)P

It is also important to note that there is usually no simple, single definition of a 'most onerous' internal pressure. A more positive internal pressure relieves loads across windward walls but increases loads across side and leeward walls. A more negative internal pressure does the opposite. But 'most onerous' can be defined for any particular action, e.g. the most onerous *uplift* on a roof will always be given by the most *positive* internal pressure.

Where there are openings in at least two sides of a building (walls or roof), the area of which is greater than 30% of that side, the building should be treated as either:

Clause 7.2.9(2)

* a free-standing canopy roof, using section 7.3, if there are at least two 'open' walls, or
* as a set of free-standing walls, using section 7.4, if there is no roof.

Section 7.3 gives values only for the canopy roof, and not for any associated wall, which gives a problem for open-sided buildings such as grandstands. The UK wind code, BS 6399-2[1] gives values of internal pressure for open-sided buildings that have a roof and at least one wall. It is reasonable to adopt these as 'non-contradictory complementary information' to make up for the deficiency of wall values.

The Note to clause 7.2.9(2) gives some recommendations on typical permeability of buildings, but permits the National Annex to proved additional information. As the airtightness of buildings varies widely across the Member States, in response to climatic variations and regulations on energy conservation, it must be expected that most National Annexes will provide additional information. The UK NA provides values for some typical construction forms in Table NA5, but these are not comprehensive.

UK NA 2.30

A 'dominant opening' is defined to occur when the area of openings in any one face is at least twice the area of openings in the other faces of a building. The Note states that this also applies to subdivisions, i.e. rooms, within a building, for determining loads on internal walls, ceilings and partitions. The internal pressure is taken to be a fraction of the external pressure at the dominant opening using the following rules:

Clause 7.2.9(4)–(5)

* Where the dominant open area is 2 times the area of other openings, the internal pressure is taken as 75% of the external pressure at the dominant opening.

- Where the dominant open area is 3 times the area of other openings, the internal pressure is taken as 90% of the external pressure at the dominant opening.

Values for proportions of opening between 2 and 3 may be linearly interpolated. This corresponds to current UK practice.

Clause 7.2.9(3)

Important warning

An opening which is normally closed, but would be dominant when open, is usually called an elective dominant opening. Clause 7.2.9(3) repeats the requirement given by Clause 2(4), and so the caveats given in section 2.3 above are also repeated here: **this is likely to lead to unsafe designs if wrongly interpreted as being sufficient on its own.**

It is normal practice to take all elective openings as being closed when assessing the ultimate limit state, but to consider them as open when assessing serviceability. In the case of a long-span roof with an elective dominant opening in the windward wall, e.g. an aircraft hangar, the net force coefficient for uplift of the roof with the door open may be more than twice the value with the door closed. The critical design case for uplift of the roof depends on the combination of the square of the probability factor $(c_{prob})^2$ (from Expression 4.2) and the partial load factor for wind γ_f (from EN 1990). For the ultimate limit case, doors closed, $c_{prob} = 1$ and $\gamma_f = 1.4$. For the service case, $(c_{prob})^2$ depends on the design risk of the service condition and $\gamma_f = 1$. If there is no provision to secure elective dominant openings in the closed position the design risk may not be decreased, $c_{prob} = 1$, and the service condition with door open will give the highest uplift.

While it is reasonable to expect that a large door, normally closed, will indeed be closed in the unlikely event of the design maximum wind storm, it is highly likely that this door will be opened in less severe wind conditions. One option is to calculate the value of c_{prob} at which the 'open' service case equals the 'closed' ultimate case, then back-calculate the corresponding annual risk p, and to decide whether this represents an acceptable risk. For buildings under construction, where the 'open' case is a temporary condition, the seasonal factor c_{season} may also be used. Provided that the elective dominant openings are secured closed in wind storms, current best practice in the UK is given by BRE Digest 436[8] – which is to apply the BS 6399-2[1] probability factor $S_p = c_{prob} = 0.85$, giving $(c_{prob})^2 = 0.72$. With these provisions, typically, the ultimate (closed) and service (open) loads for the critical design cases are nearly equal.

Important warning

As the liability devolves to the user of the building, it is necessary to establish a protocol for closing up elective dominant openings before the wind speed increases to a value at which the service limit state is exceeded and to ensure that the building owner understands his/her responsibilities in this respect.

Clause 7.2.9(6)

For all buildings without dominant openings, the internal pressure is given in a chart, Fig. 7.13, depending on the proportions of the building. This chart is reproduced with some additional annotations as Fig. D7.2 below. The chart requires the proportion parameter, μ, defined as:

$$\mu = \frac{\sum \text{area of openings where } c_{pe} \leq 0}{\sum \text{area of all openings}} \qquad (D7.1)$$

In general, the area of openings where $c_{pe} \leq 0$ will be all openings on faces other than the windward wall and the windward roof pitch when this is steep. The annotations in Fig. D7.2 show the corresponding range of c_{pi} values in BS 6399-2.[1] Values of μ are shown or some proportions of building plans, assuming that each wall is equally porous (and the roof is impermeable). The smallest typical value is $\mu = 0.5$ for a slab building with porous front and rear faces and impermeable side walls and roof, resulting in a small positive internal pressure coefficient. Smaller values of μ, so more positive internal pressures, are only possible when the windward wall is more porous than the other walls. The internal pressure coefficient for the majority of typical buildings will be slightly negative.

Current UK practice for the internal pressure coefficient of buildings is to take $c_{pi} = -0.3$ for the typical case of 'four walls equally permeable' ($\mu = 0.75$) and $+0.2$ for the less typical case of 'windward and leeward walls equally permeable, side walls impermeable' ($\mu = 0.5$). The EN rules for internal pressure coefficient tally quite well with current UK practice,[1] but are more precise.

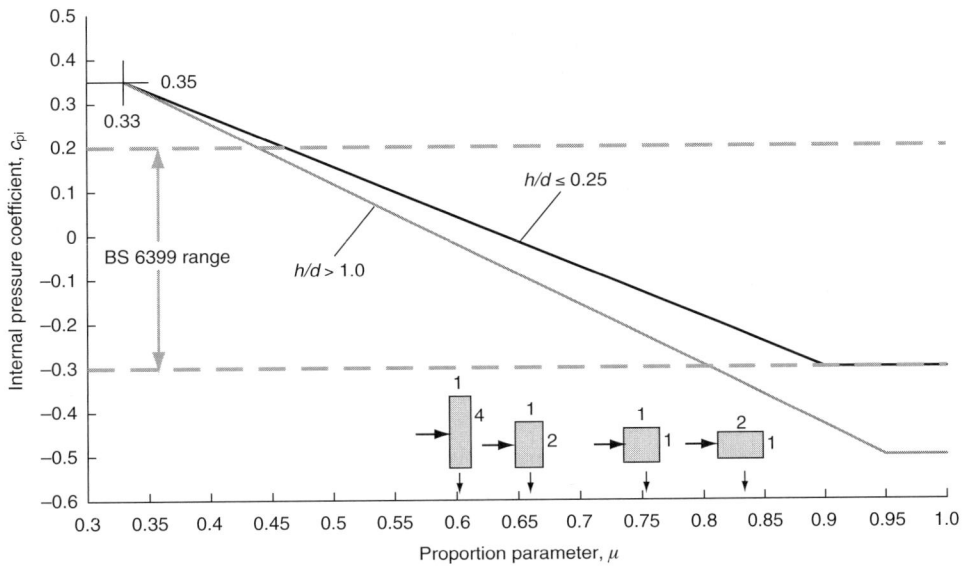

Fig. D7.2. Internal pressure coefficients for uniformly distributed openings

In all cases, the reference height z_e for internal pressures is equal to the reference height for the corresponding external pressures on faces that contribute to the internal pressure. Where there are several contributing openings, which includes buildings without dominant openings, the largest reference height should be used. This implies that, when a building is stepped in height and has a common internal volume, the reference height should be taken at the top of the taller part. However, if the taller part is small, the rest of the building may control the internal pressure. The largest value of internal pressure, positive or negative, is given by this rule but may not be the most onerous value, depending on the sign of the corresponding external pressure, so the rule is not necessarily conservative.

Clause 7.2.9(7)

Internal pressure coefficients are also given for open-topped silos and chimneys and for vented tanks.

Clause 7.2.9(8)

7.2.10. Pressures on walls and roofs with more than one skin
Typical examples of walls and roofs comprised of more than one skin are:

Clause 7.2.10(1)

- cavity walls
- rain-screen cladding
- bitumen-felt built-up roofs
- tiles or slates over sarking or close-boarded roofs.

The overall pressure difference across all layers may be determined directly from the external and internal pressures. Where the layers act as a composite structure, this may be sufficient information to execute the design. Where the pressure difference across any single layer is needed, the value depends on the relative porosity of each layer.

Clause 7.2.10(2)

A skin is defined as impermeable when the permeability μ, defined as the ratio of the total area of openings to the total area of skin, is less than 0.1%. **This is not the same definition as in equation (D7.1) above – the symbol μ has been reused here for a different purpose.** Note that, if two skins are impermeable, the pressure of the volume of air trapped between them will depend on the atmospheric pressure and temperature through the gas equation. However, when the volume is small compared with the area, as is the case for double-glazed window panels, the net pressure difference can be assumed to be shared equally between the two skins.

Important warning

If only one skin is permeable, the impermeable skin should be assumed to bear the full pressure difference, $c_{p,net}$. The mean pressure difference across the permeable skin will approximate zero, but will fluctuate in time and may bear an instantaneous load up to

Clause 7.2.10(3)

about $c_{p,net}/3$. This is the proportion assumed to apply to rain-screen cladding, flush-laid pavers and insulation boards[17] in current UK design practice. However, EN 1991-1-4 assumes that this applies only to suctions ('*underpressure*') and recommends that the proportion for positive pressures ('*overpressure*') should be taken as 2/3 of $c_{p,net}$. This leads to a number of recommended options given in a Note, which apply:

- when the extremities of the layer between skins are airtight, as defined by Fig. 7.14 of the EN. In practice, this means that the cavity must be effectively sealed along all edges, i.e. at the corners and eaves for a cavity wall
- when the free distance between skins is less than 100 mm.

The recommended rules are summarized as follows:

- The pressure difference across the most rigid skin may conservatively be taken as the full pressure difference $w_{net} = w_e - w_i$.
- Take $w_{net} = w_e - w_i$ across an impermeable inside skin, and $w_{net} = (w_e - w_i)/3$ for suction, and $w_{net} = 2 \cdot (w_e - w_i)/3$ for pressure across a uniformly permeable outside skin.
- Take $w_{net} = w_e - w_i$ across an impermeable outside skin, and $w_{net} = w_i/3$ across a uniformly permeable inside skin. Here, the load on the inside skin comes only from fluctuations of the internal pressure.
- When both outside and inside skins are impermeable and the inside skin is more rigid, take $w_{net} = w_e$ across the outside skin and $w_{net} = w_e - w_i$ across the inside skin.

EN 1991-1-4 gives these rules in terms of the pressure coefficients c_{pe} and c_{pi}. As it is possible to have a different reference height z_e for the external and internal pressures, e.g. for a stepped-height building, it is better practice to express the rule in terms of the resulting pressures, w_e and w_i.

Note that the sum of the loads across two skins can never exceed the net pressure $w_{net} = w_e - w_i$, implying that the design values for individual skins do not apply simultaneously. The partitioning procedure described in section 7.2.2 above is also applicable here.

To become normative, these rules must be adopted by the National Annex, otherwise they must be replaced by other rules. The UK NA allows the rules to be used except for the case of tiled or slated roofs, which should be designed to BS 5534,[18] and masonry cavity walls, which should be designed to EN 1996-1.[19]

Note that the scope of these rules excludes an increasingly common trend of installing a glazed 'wind screen' at a sufficient distance away from a building façade to allow windows to be opened for ventilation into the protected volume between the façade and the screen. The exclusion occurs because this gap is typically greater than 100 mm and the corners are often unsealed. This deficiency will need to be addressed in later revisions but in the meantime designers who wish to use screens to promote the use of natural ventilation in taller buildings will need to seek specialist advice.

7.3. Canopy roofs

These are defined as roofs of structures that do not have permanent walls, so that zones are defined for net coefficients on roof areas only and no values for walls are given. Canopy roofs include such structures as petrol stations or Dutch barns. Unfortunately, the user is directed to the section from clause 7.2.9(2) in the case of a building with two open sides and two walls – see section 7.2.9, earlier – so the user will need to refer to a source of non-contradictory complementary information e.g. 15 or seek specialist advice. Canopies over entrances to buildings are also excluded.

The pressure underneath a canopy roof depends on how easily the wind can flow underneath and, hence, on the degree of blockage. The blockage ratio φ is defined as the ratio of the area of obstructions under the canopy to the total cross-sectional area under the canopy viewed in elevation, so that $\varphi = 0$ represents an empty canopy and $\varphi = 1$ represents a canopy blocked from ground to eaves over the full width.

Values of the net pressure coefficient across the roof $c_{p,net}$ are given for $\varphi = 0$ and $\varphi = 1$ and values for intermediate blockage ratios may be interpolated. Positive values indicate a net downwards force and negative values indicate a net upwards force. Upwind of the position of maximum blockage φ the corresponding values should be used, but downwind of the position of maximum blockage the values for $\varphi = 0$ should be used. If contents are stacked level, e.g. as in a Dutch barn, it would be sensible to use the more onerous loading.

Clause 7.3(3)–(5)

Specific rules for asymmetry of loading are given in clause 7.3(6) and illustrated by Figs 7.16 and 7.17, which also define the reference height z_e as the eaves height (highest eaves for monopitch canopies). Friction forces need to be considered (see section 7.5 below), bearing in mind that both the upper and the lower surfaces of the canopy will be swept by the wind. Reduction factors are given for the second and subsequent bays of multibay canopies.

Clause 7.3(6)–(9)

Overall, the procedures and values for canopy roofs correspond to current UK practice.

7.4. Free-standing walls, parapets, fences and signboards

This section does not apply to parapets and noise barriers on bridges, which are dealt with in Chapter 8. Values are given as net pressure across the wall, etc. This precludes the use of the 'division by parts rule' (see section 7.2.2 above) unless assumptions are made on how the net pressure is divided between windward and leeward faces.

Clause 7.4(1)

The solidity ratio for walls and fences shares the same symbol φ with the blockage ratio for free-standing canopies and is defined here as the solid area divided by the area of the envelope, hence $\varphi = 0.8$ for a wall that is 80% solid and 20% open. It is assumed that the openings are reasonably uniformly distributed. If this is not the case – for example a solid wall with a fence on top if it – the structure should be divided into parts that are reasonably uniform and assessed using φ for each part. Note, however, that the reference height z_e for each part will remain the full height of the structure.

Any porous wall or fence with $\varphi \leq 0.8$ should be treated as a plane lattice using the rules described in section 7.11 below. Many typical fences will fall into this category, but the rules for plane lattice frames do not include the effect of shelter by other upwind frames (fences). The Guide to BS 6399-2[20] contains shelter rules that may be considered as non-contradictory complementary information.

7.4.1. Free-standing walls and parapets

The essential difference between a free-standing wall and a parapet is that a free-standing wall stands on the ground while a parapet stands on top of a building. The rules may be applied to any other free-standing wall between these two limits, e.g. a boundary wall along a step change in ground level.

Clause 7.4.1(2)

In all these cases, the reference height z_e is the height from the ground to the top of the wall, etc., on the windward side.

Values of pressure coefficient are presented for solidity ratios of $\varphi = 1.0$ (solid) and $\varphi = 0.8$, for zones defined in terms of wall heights from a free end or return corner of the wall, etc. Loads are highest close to a free end and the maximum load occurs when the wind is skewed about $45°$ onto the free end. These high end loads increase in value as the wall, etc., becomes longer. Return corners reduce these loads, but must extend at least one wall-height from the corner to be effective.

Clause 7.4.1(1)

The Note requires the National Annex to make these values normative or to replace them with other values. As these values are very similar to current UK practice, the UK NA adopts them without change.

UK NA 2.31

The provenance of these values is very good, having first been measured using wind tunnel models at two independent laboratories and later confirmed by measurements on full-scale walls in the natural wind. The increase in end loads with overall wall length was a surprising finding but, very recently, this has also been shown to apply to long buildings (e.g. terraced houses), but this new finding is not reflected in EN 1991-1-4.

7.4.2. Shelter factors for walls and fences

Clause 7.4.2(1)

EN 1991-1-4 gives an allowance for shelter on free-standing walls and fences caused by other upwind walls and fences, provided that these sheltering walls and fences are at least as tall as the wall or fence being considered. Nowhere else in the code is there a similar provision that allows the user to take account of the direct shelter caused by one structure on another, because there is always the possibility that this shelter may be lost through demolition. The allowance for displacement height h_{dis} already given in section 4.3.5 above relies on a number of sheltering obstructions and is limited in value, so that demolition of any one would not erode the design risk.

Direct shelter is permitted in this instance because:

- typical boundary fences and walls are designed as integral components of suburban developments
- many would not perform satisfactorily without the benefit of this shelter, and
- failure of a fence (usually by overturning) does not usually pose a significant risk of injury.

However, where there is a significant risk of injury – for example a masonry boundary wall that may fall across a right of way – the user should consider the associated risk to the public if it is not possible to ensure that the shelter will always remain in place. Clearly, walls and fences around the outer boundary of a development will need to bear the full wind load in winds from the direction of full exposure.

Values of shelter factor ψ_s are given only for the solidity ratios of $\varphi = 1.0$ (solid) and $\varphi = 0.8$ for the upwind, sheltering wall, i.e. only for shelter by walls and fences covered by this section. This implies that more open fences, designed using section 7.11 below, provide no shelter at all. The guide to BS 6399-2[20] provides values of shelter factor for the full range of solidity $0 \geq \varphi \geq 1$, and these may be taken as 'non-contradictory complementary information'.

Clause 7.4.2(2)

The shelter factor may not be applied to the end zones within one height h from the end of the wall, i.e. not for zone 'A' and part of zone 'B'. This is to allow for wind flowing around the upwind end of the sheltering wall, i.e. assuming that both walls have similar ends. However, if the sheltering wall is significantly longer than the loaded wall, the end zone would also be significantly sheltered, but this must be discounted.

7.4.3. Signboards

Clause 7.4.3(1)–(3)

Signboards are defined here as either:

- having a gap between the bottom of the signboard and the ground of at least $h/4$ through which the wind may flow, or
- being taller than they are wide, $b/h \leq 1$,

when the single value of $c_f = 1.80$ applies. Otherwise, wider signboards that reach the ground, or have a smaller gap underneath, must be designed as boundary walls using section 7.4.1 above.

Asymmetry of load is specifically addressed by using a horizontal displacement of the centre of force, e, from the centre of the signboard. There is National Choice for the recommended value of $e = \pm 0.25b$, and this value is adopted by the UK NA, except for road signs as defined in BS EN 12899-1.

UK NA 2.33

Signboards mounted on slender legs or poles are susceptible to divergence or stall flutter instabilities and should be checked using the rules in Annex E. Divergence is a static torsional instability which occurs when the wind moment increases faster than the torsional resistance. As the wind moment is proportional to the square of wind speed and the torsional resistance is constant, divergence occurs when a critical wind speed value is exceeded and the structure rotates suddenly to catastrophic failure. Stall flutter (strictly, 'stall-limited divergence') is a special case of divergence, where the rotation is limited by a sudden decrease in aerodynamic moment when the flow separates and 'stalls'. The structure recovers and then rotates in the opposite direction until the flow stalls again. This oscillation is repeated until the wind drops below the critical wind speed or the structure fails by fatigue. Stall flutter is also commonly called 'stop-sign flutter' because it frequently occurs with the octagonal STOP signs in the USA which are usually mounted on channel section stakes with minimal torsional stiffness.

7.5. Friction coefficients

'Friction' is the term used to describe the accumulated tangential shear force F_{fr} caused by flow skimming surfaces parallel, or nearly parallel, to the wind direction. This friction force is significant only where the structure is long in the wind direction, or has only a small area normal to the wind. The user is referred to clause 5.3(3) of the EN for the cases to be considered. This clause gives the required expression for F_{fr}, and refers back to clause 7.5 for the definition of the 'area of external surface parallel to the wind'. It is useful to note that clause 5.3(4) allows the effect of friction to be disregarded when the total area of surface parallel to the wind is less than 4 times the total area normal to the wind (see section 5.3 above).

Clause 7.5(1)

Table 7.10 gives the values of the friction coefficients to be used, which depend on the roughness of the surface, but are significantly smaller in value than the normal pressure coefficients. Although the table defines the same three values currently in BS 6399-2,[1] each applies to a different range of surface:

Clause 7.5(2)

- The smallest coefficient for 'smooth', $c_{fr} = 0.01$, which in BS 6399-2 applies to all surfaces without corrugations and ribs across the wind direction, now applies only to smooth surfaces, such as steel or smooth concrete.
- The middle coefficient for 'rough', $c_{fr} = 0.02$, which in BS 6399-2 applies to surfaces with corrugations across the wind direction, now applies to rough surfaces, such as rough concrete and 'tar boards'.
- The largest coefficient for 'very rough', $c_{fr} = 0.04$, which in BS 6399-2 applies only to surfaces that are ribbed across the wind direction, now applies to all surfaces that have ripples, ribs or folds, irrespective of the wind direction. All corrugated, profiled or ribbed cladding panels fall into this category.

These provisions are clearly more onerous than current UK practice, but are unlikely to have a large effect on design owing to the (usually) small contribution from friction.

Although the EN does not distinguish the wind direction when dealing with ribs and corrugations, it is reasonable for the user to do so and to interpret surfaces with ribs or corrugations parallel to the wind direction as being 'smooth'.

Figure 7.22 defines the reference areas over which friction should be considered to act for three datum cases:

Clause 7.5(3)

- A horizontal plate-like structure supported on legs, such as a canopy over a petrol station, where friction acts on the whole upper and lower surface in all wind directions.
- A vertical plate-like structure, such as a free-standing wall, where friction acts on the whole of both sides, but only when the wind is parallel to the wall.
- An enclosed building, where friction acts on the downwind parts of the sides and roof located more than $2b$ or $4h$, whichever is the smaller, from the upwind verge or corner, and only with the wind parallel to the ridge line. (EN 1991-1-4 uses the term 'eaves' to mean verge for the case illustrated in Fig. 7.22.) The upwind region that is discounted represents the zone of flow separation, where the normal pressures are suctions and the flow at the surface may be reversed.

The user should not assume that these are the only cases that clause 5.3(3) requires to be considered. For example:

- A pitched canopy roof with no blockage should be treated the same as the first case, above, when the wind is parallel to the ridge. However, when fully blocked, this reverts to the third case, with no flow under the canopy and a flow separation region behind the verge which is discounted.

- It may be assumed that there is no friction acting on pitched roofs when the wind is normal to the ridge or eaves. However, a flat-roofed building will experience friction on the defined downwind region from all wind directions because there is no ridge.
- In the case of a multi-bay building, with wind normal to the ridges, there will be no region of friction on the roof, but there will be friction on the side walls if they are sufficiently long.
- A building with no closed gables at either end will allow wind to blow through it, such that the 'external area parallel to the wind' comprises the top and bottom surfaces of the roof, inside and outside surfaces of both side walls.

Clause 7.5(4)   The reference height z_{ref} is always the height of the structure above ground h. Current UK practice would allow the eaves height to be used as reference for the friction on side walls, but Fig. 7.22 is explicit in not allowing this interpretation.

7.6. Structural elements with rectangular sections

Clause 7.6(1)   The force coefficient c_f for structural elements of rectangular section for wind normal to a face is given by expression 7.9 which uses:

- a basic value $c_{f,0}$ which corresponds to sharp corners and infinite length
- a reduction factor ψ_r which accounts for rounding of the corners
- an end-effect factor ψ_λ which accounts for the length of the element between free ends. This factor is common to all the following forms of structural element and is defined after these forms in section 7.13 below.

The basic value $c_{f,0}$ is given by Fig. 7.23 and depends on the cross-sectional shape d/b: starting at $c_{f,0} = 2.0$ for a thin plate normal to the wind, $d/b = 0$; rising to $c_{f,0} = 2.4$ for the critical proportion $d/b = 0.7$; then falling to $c_{f,0} = 0.9$ for $d/b > 10$.

UK NA 2.34   Note 1 permits National Choice for reduction factor ψ_r, but the UK NAD adopts the recommended upper-bound values given in Fig. 7.24. Note 2 permits the use of the reduction factor ψ_r for tall buildings with rounded corners.

Clause 7.6(2)   The reference area is the area of the windward face of the element, $A_{ref} = \lambda \cdot b$, where λ is defined locally as the length of element being considered. The reference height z_{ref} is the maximum height above ground of the section being considered. Note that when these rules are applied to buildings, the rules in clause 7.2.2 set limits to the way the building can be divided into sections.

Clause 7.6(3)   Thin plate-like sections, $b/d < 0.2$, may act like wings at certain skew angles of the wind, giving higher 'lift' forces normal to the plate. This clause recommends that c_f is increased by 25% to cover this eventuality.

7.7. Structural elements with sharp-edged section

Clause 7.7(1)   Sharp-edged structural sections of angle, T-section, I-section and channel forms are covered by this section. The force coefficient is defined as the base coefficient $c_{f,0}$ multiplied by the common end-effect factor ψ_λ given in section 7.13.

UK NA 2.35   The recommended value of the base coefficient is $c_{f,0} = 2.0$. but is open to National Choice. The recommended value is almost universally accepted in international codes, so most NAs are likely to use it. The UK NA uses the recommended value.

Clause 7.7(2)–(3)   The reference area for in-wind (drag) forces is $A_{ref,x} = \lambda \cdot b$ and for cross-wind (lift) forces is $A_{ref,y} = \lambda \cdot d$, where λ is the length of element being considered, i.e. the normal area of the element in each orthogonal direction. The reference height z_{ref} is the maximum height above ground of the section being considered.

7.8. Structural elements with polygonal section

Clause 7.8(1)

This section covers regular polygonal sections, starting with pentagons and proceeding with increasing number of sides towards the limit of a circular cylinder. The force coefficient is again defined as the base coefficient $c_{f,0}$ multiplied by the common end-effect factor ψ_λ given in section 7.13.

The recommended value of the base coefficient $c_{f,0}$ is given for polygons of 5, 6, 8, 10, 12 and 16–18 sides in Table 7.11, taken from a number of disparate sources, but is open to National Choice. The recommended values are expected to give a safe upper bound. Reynolds number dependence becomes significant as the number of sides increases; however, the values do not completely cover this effect, i.e. it is included in Table 7.11 for 8 and 12 sides, but is missing for 10 sides. The definition of Reynolds number is given in the following section 7.9.1. It is reasonable to assume that values for the missing numbers of sides, 7, 9, etc., may be interpolated between the values given, although the EN includes no note to this effect. It also follows that the Reynolds number dependence for 10 sides might be interpolated. The UK NA adopts the recommended values.

UK NA 2.36
Clause 7.8(2)

The drag force on polygonal-plan buildings with $h/d > 5$ may also be determined using this section. There are no corresponding lift forces for these regular polygonal sections.

Clause 7.8(3)–(4)

The reference area for in-wind (drag) forces is $A_{ref,x} = \lambda \cdot b$, where λ is the length of element being considered and breadth b is locally defined in Fig. 7.26 as the diameter of the smallest enclosing circle. The reference height z_{ref} is the maximum height above ground of the section being considered.

7.9. Circular cylinders

7.9.1. External pressure coefficients

Clause 7.9.1(1)

Pressure coefficients of circular cylinders depend on the wind speed and the diameter through the Reynolds number defined by Expression (7.15) as:

$$\text{Re} = \frac{b \cdot v(z_e)}{\nu} = \frac{b \cdot v(z_e)}{15 \cdot 10^{-6}} \qquad (D7.2)$$

where ν is the kinematic viscosity of air (in m²/s), b is the diameter (in m) and $v(z_e)$ is the peak wind velocity (in m/s) at the reference height. (This is the one expression in the EN where the Greek 'nu' symbol ν and Roman 'vee' symbol v are used together – see section 1.7 above.)

Clause 7.9.1(2)–(4)

The force coefficient is defined as the base coefficient $c_{p,0}$ multiplied by an end-effect factor $\psi_{\lambda\alpha}$. The base coefficient $c_{p,0}$ is given in Fig. 7.27 for various Re as a function of the angle α around a smooth cylinder, measured from the front. The end-effect factor $\psi_{\lambda\alpha}$ is given by Expression (7.17) as a function of α and the common end-effect factor ψ_λ given in section 7.13 below.

The determination of c_{pe} for cylinders is not simple, given the number and complexity of the component factors to determine Re dependence and end effects, but the outcome may be summarized as follows:

- The lobe of positive pressure at the front of the cylinder is independent of Re and of end effects.
- The zone of relatively steady suctions at the rear is dependent on Re and the end effects are given by the common end-effect factor ψ_λ only.
- The lobes of high suctions either side of the cylinder are the most dependent on both Re and on end-effect, requiring the end-effect factor $\psi_{\lambda\alpha}$ given by Expression (7.17).

It is an unfortunate fact of the physics that the highest suctions in the side lobes are the most complex to determine. However, the end-effect factor is always less than unity, so that the base coefficients in Fig. 7.27 of the EN provide safe upper-bound values.

Clause 7.9.1(5)–(6)

The reference area $A_{ref} = \lambda \cdot b$, where λ is the length of element being considered and b is the diameter. The reference height z_{ref} is the maximum height above ground of the section being considered.

7.9.2. Force coefficients

The force coefficient for circular cylinders is defined as the base coefficient $c_{f,0}$ multiplied by the common end-effect factor ψ_λ given in section 7.13 below. Values of $c_{f,0}$ for cylinders are given by Fig. 7.28, as a function of Reynolds number Re and surface roughness k. Figure 7.28 includes expressions for the segments of each graph curve to allow the values to be calculated directly. Values of surface roughness for a range of finishes are given in Table 7.13. The exception to these rules are stranded cables, where $c_{f,0} = 1.2$ for all Re.

The reference area $A_{ref} = \lambda \cdot b$, where λ is the length of element being considered and b is the diameter. The reference height z_{ref} is the maximum height above ground of the section being considered.

The EN requires the user to seek specialist advice when cylinders are mounted close to a plane surface (e.g. horizontal cylinders close to the ground) when the gap between the cylinder and the surface is less than $1.5b$. This will be necessary for a number of common situations, such as over-ground pipelines and external down-pipes on buildings. Non-contradictory complementary guidance is available in [Reference 15] which, in summary, shows that the drag force coefficient increases to a maximum of $c_f = 1.45$ at a gap of $0.4b$, as the ground plane is approached and falls to $c_f = 0.8$ touching the ground plane. A sideways lift force, acting away from the plane, is insignificant with a gap of b, is $c_f = 0.15$ at a gap of $0.4b$, and rises to a maximum of $c_f = 0.60$ touching the plane.

7.9.3. Force coefficients for vertical cylinders in a row arrangement

This situation is common for the arrangement of stacks or flues rising from the roof of a building. The force coefficient is determined in exactly the same way as for an isolated cylinder, with the addition of an enhancement factor κ which is given in Table 7.14 and varies in value from $\kappa = 1.0$ (no effect) when the centres of the cylinders are more than 30 diameters apart to $\kappa = 1.15$ (maximum effect) when the centres are 3.5 diameters apart. A linear interpolation expression is given between these limits.

7.10. Spheres

The along-wind force coefficient $c_{f,x}$ for a sphere is given in Fig. 7.30 for a range of Reynolds number and surface roughness k and are applicable when there is a gap of more than half a diameter between the sphere and any plane surface. When the gap is less than half a diameter, the value of $c_{f,x}$ is increased by a factor of 1.6. These values are subject to National Choice and have been adopted by the UK.

The vertical force coefficient $c_{f,z}$ for a sphere is given as $c_{f,x} = 0$ when there is a gap of more than half a diameter between the sphere and the ground surface and $c_{f,x} = 0.60$ when the gap is less.

In both these cases the reference area A_{ref} is the projected area of the sphere $A_{ref} = \pi \cdot b^2/4$ and the reference height is the centre of the sphere.

7.11. Lattice structures and scaffoldings

This section covers only single plane lattice frames, and three-boom and four-boom lattice trusses. In general, the procedure for lattice structures is identical to that for the preceding structural elements, using a base force coefficient and the common end-effect factor, except for variations in the definitions of reference area. However, the value of the base force coefficient $c_{f,0}$ now varies with the solidity ratio φ of the face of the frame or truss.

It should be noted that the force coefficients only apply to open lattice frames consisting of one type of member (e.g. all sharp-edged or all circular), and not containing any ancillaries such as ladders and cables. Procedures to derive force coefficients for such configurations are given in prEN 1993-3-1 Annex B[21] covering lattice towers and masts.

Note 2 allows National Choice on a reduction in the loads acting on the scaffolding when it is sheltered by an adjacent solid building, recommending the value in EN 12811.[22] The UK

NA adopts this recommendation, but also references published information on the reductions in overall loads on arrays of frames due to mutual sheltering. This information may be taken as non-contradictory complementary information (NCCI).

7.12. Flags

Rules are given for the calculation of the forces imposed by 'fixed' (e.g. banners) and 'free' (e.g. typical fluttering) flags on their supports. The provenance of the equation for free flags is unknown, but the data in earlier codes of practice have been traced back to tests on advertising banners towed by aircraft which were determined by testing in response to a series of fatal accidents in the 1920s.

Clause 7.12(1)–(2)

A simple expedient to avoid overloading the supporting structure is to insert a 'fuse' – that is, a breakable element – in the fixing. The 'fuse' should snap, allowing the flag to blow away before the structure becomes overloaded. Wind and safety screens attached to scaffolding around buildings are sometimes attached using breakable elastic ties.

7.13. Effective slenderness λ and end-effect factor ψ_λ

The force coefficients $c_{f,0}$ given in sections 7.6 to 7.12 above are based on infinitely long sections. The end-effect factor accounts for reduced force caused by wind flow around the ends of a finite section. There is National Choice in the value of the factor, ψ_λ, but the UK NA adopts the recommended values given in Fig. 7.36.

Clause 7.13(1)–(3)
UK NA 2.39

The factor depends on the effective slenderness λ and there is also National Choice in this definition, with the recommended definitions in Table 7.16. The UK NA replaces these definitions, discarding No. 3 and simplifying the remainder. (There is a danger that 'No. 3' in Table 7.16 may be misconstrued as applying to a free-standing wall adjacent to a building, despite Table 7.16 being clearly captioned as applying to 'sections'.)

UK NA 2.40

CHAPTER 8

Wind actions on bridges

This chapter is concerned with defining the rules for bridges, which are treated as an exceptional case in *Section 8*. In effect, *Section 7* is a code for buildings and *Section 8* is a code for bridges, both using common design wind speed information and application models from the previous sections. The material in this chapter is covered in the following clauses:

• General	*Clause 8.1*
• Choice of the response calculation procedure	*Clause 8.2*
• Force coefficients	*Clause 8.3*
• Bridge piers	*Clause 8.4*

8.1. General

The rules for bridges in EN 1991-1-4 are restricted to self-supporting bridges (less than 200 m long – see section 1.1 above) with constant depth and cross-section, where:

Clause 8.1(1)

- the bridge deck is stiffened by longitudinal beams or box-beams
- the bridge deck is stiffened by trusses or plates above or below the road deck, or
- the bridge has a box-girder deck.

Note 1 allows rules for most other forms of bridge, including arch, suspension, cable-stayed, to be included in the National Annex. As most of these bridges will be designed using specialist guidance, it is unlikely that many NAs will adopt rules for these bridges. The UK NA does not give any additional guidance on overall structural loads. However, the UK NA does allow elements of such bridges to be designed using the relevant clauses of the EN, with some modifications by the UK NA.

UK NA 2.41

Note 2 permits the NA to define the angle of the wind to the deck in both vertical and horizontal planes. The UK NA adopts the recommended definition.

UK NA 2.42

The calculation of wind forces on bridges is broken down into two steps:

Clause 8.1(2)

- forces exerted on the deck (clauses 8.2 and 8.3)
- forces on the supporting piers.

In combining these loads, the forces should always be considered to act simultaneously when they increase the overall load, i.e. when they are 'unfavourable'.

The coordinate axes for forces on bridges are defined as the x-direction normal to the span in the plane of the deck, y-direction parallel to the span in the plane of the deck, and z-direction normal to the deck. The most onerous forces in the x and y directions are not simultaneous because they are caused by wind blowing from different, usually orthogonal, directions – normal to the span for x and along the span for y. However, as the maximum force in the z direction may occur in any wind direction, it should be considered to act simultaneously with x or y.

Clause 8.1(3)

Clause 8.1(3),
Note

Clause 8.1(4)

UK NA 2.43

The notation of the dimensions used for bridges differs from the definitions of section 1.6. While the general definition is given in the Note, this notation may need to be adjusted for specific bridge forms.

When considering the simultaneous action of road traffic with wind, it is reasonable to limit the design wind speed to a value that represents the highest wind speed that traffic would be able to use on the bridge without incidents of overturning on the approach to the bridge. The default value of the basic wind speed is defined as $v_{b,0}^* = 23$ m/s, which may be changed in the NA. The UK NA requires $v_{b,0}^* = v_{b,0}$ (i.e. the fundamental basic wind velocity appropriate for the site) to be used in calculations of peak velocity pressure $q_p(z)$, but limits the resulting value to a maximum value of $q_p(z) = 750$ Pa which corresponds to a mean wind speed of 23 m/s (as recommended in the EN) or gust speed of 35 m/s.

Clause 8.15
UK NA 2.44

Similarly, when considering rail traffic simultaneously with wind, a somewhat higher design wind speed is appropriate, because railway rolling stock is more stable in wind than road vehicles. The default value of the basic wind speed is defined as $v_{b,0}^{**} = 25$ m/s, which may be changed in the NA. Again, the UK NA requires $v_{b,0}^* = v_{b,0}$ (i.e. the fundamental basic wind velocity appropriate for the site) to be used in calculations of peak velocity pressure $q_p(z)$, but now limits the resulting value to a maximum value of $q_p(z) = 890$ Pa which corresponds to a mean wind speed of 25 m/s (as recommended in the EN) or gust speed of 38 m/s.

8.2. Choice of the response calculation procedure

Clause 8.2(1)

An assessment is required to determine if the bridge is dynamically sensitive and so requires an assessment of $c_s c_d$. The Notes allow criteria and procedures to be given in the NA, but recommends $c_s c_d = 1.0$ when no assessment is needed and suggests that no assessment is needed for '*normal road and rail bridge decks of less than 40 m span*' of the deck types shown in Fig. 8.1.

UK NA 2.45

The UK NA includes two procedures derived from BD 49/01 *Design Rules for Aerodynamic Effects on Bridges,*[23] which comply with current UK practice. The first determines whether dynamic magnification effects are significant in the in-wind direction, or can be ignored. The second categorizes the aeroelastic stability of the bridge using an aerodynamic susceptibility parameter P_b.

- If $P_b < 0.04$ and the bridge is built of '*normal construction*', i.e. built of steel, aluminium and/or timber, and with a shape generally covered by Fig. 8.1, then all aerodynamic excitation effects may be considered insignificant.
- If $0.04 \leq P_b \leq 1.00$ the bridge is deemed to be within the scope of the EN and the UK NA procedures.
- If $P_b > 1.00$ the bridge is deemed to be very susceptible to aerodynamic excitation and wind tunnel testing is mandatory.

Dynamic magnification is expected to be insignificant for most typical short-span (<12 m) highway and railway bridges. Long-span bridges (>250 m), particularly suspension or cable-stayed bridges, are expected to fall into the third category, but are outside the scope of the EN (see section 1.1). This leaves a significant proportion of intermediate-span bridges for which the EN and UK NA procedures are appropriate. However, the UK NA stipulates that '*covered footbridges, cable supported bridges and other structures where any of the parameters b, L, or n_{1b} cannot be accurately defined*' must be placed into the third category requiring wind tunnel tests. This stipulation is an unfortunate consequence of new restrictions introduced in the 2001 revision of BD 49.[23] The previous version, BD 49/93, was known to be grossly conservative in its treatment of footbridges, but the latest amendment has made this situation even worse. Recent trends for innovative cable-stayed footbridges are particularly affected. While wind tunnel testing may be justified for major river crossings (such as the infamous 'wobbly' Millennium Footbridge), the expense of such testing militates against innovative solutions for footbridges across minor streams or roads.

8.3. Force coefficients

This section gives force coefficients for the main bridge deck only – pressure coefficients are not used at all for bridges. Force coefficients for the design of parapets and gantries on bridges may be given in the National Annex, but the Note recommends using the values in section 7.4 of the EN. The UK NA directs the user to the appropriate clauses.

Clause 8.3(1)
UK NA 2.46

8.3.1. Force coefficients in the x-direction – general method

The force in the *x*-direction is normally referred to as the 'drag normal to the deck'. The *x* force coefficient is the value without free-end flow, $c_{fx} = c_{fx,0}$, because the road/track way is continuous to the end abutments (except, perhaps, during construction). The value may be taken as $c_{fx,0} = 1.3$ or from Fig. 8.3 when the angle of inclination of the wind to the deck is less than $10°$.

Clause 8.3.1(1),
Notes 1–3

Inclined flow will occur when the bridge deck is not horizontal, but could also occur if there is a significant slope of the terrain under the bridge. Special studies, e.g. wind tunnel tests, are required if the angle of inclination exceeds $10°$.

When the windward face of the bridge deck is inclined to the vertical, as is quite common for box-girder decks, the *x*-force coefficient may be reduced by 0.5% per degree of this inclination, limited to a maximum 30% reduction. This reduction is not applicable to the simplified method of section 8.3.2 of the EN unless specifically permitted by the NA. The UK NA does not permit this reduction.

Clause 8.3.1(2)
UK NA 2.47

When the bridge deck is '*sloped transversely*', i.e. when the deck is not horizontal or there is a significant camber across the deck (such that the resolved depth in the wind direction is greater than the actual depth *d*), the *x*-force coefficient may be increased by 0.3% per degree of inclination, limited to a maximum 25% increase.

Clause 8.3.1(3)

The reference area $A_{ref,x}$ for wind load combinations without traffic loads is obtained by summing the areas of the windward faces of the bridge deck components.

Clause 8.3.1(4)

For decks with solid plate beams and box-girders this is effectively the solid area of the whole deck as seen in projection. An exception to this general rule accounts for open parapets, railings and safety barriers, where each is taken to contribute an area of $0.3\,m^2$ per metre run. Example combinations of parapet and barrier are given in Table 8.1.

For decks with trussed girders, this is effectively the solid area in projection summed for all parts of the deck, including all trusses. This assumes that the downwind trusses are not sheltered by the upwind trusses. It is possible that the total area of all trusses exceeds the envelope area of the bridge, i.e. is greater than the area of the whole deck seen in projection with the trusses assumed to be solid. This equivalent 'solid' bridge represents a reasonable, but not absolute, upper limit to the drag of the bridge, so the EN limits the total area of all trusses to the value of the envelope area.

For decks with several main girders, the reference area during construction and before the top deck is placed is taken as the projected area of two main girders. There would normally be more than two main girders, so this allows for some shelter between multiple beams placed one behind the other, where the absence of the top deck allows some wind to penetrate behind the upwind girder.

When considering load combinations of wind with traffic, allowance must be made for the additional wind loads acting on the traffic itself. The reference area for road bridges is increased by $2\,m^2$ per metre run on the most unfavourable length of the bridge, independently of the location of the vertical traffic loads. This may lead to the notional, but irrational, situation where traffic applies horizontal wind loads where there are no vertical traffic loads and vice versa. The reference area for rail bridges is increased by $4\,m^2$ per metre run for the total length of the bridge, implying a 4 m-high train occupying the whole bridge.

Clause 8.3.1(5)

The reference height above ground, z_e, may be taken as the distance from the lowest point of the ground under the bridge to the centre of the bridge deck structure, disregarding the deck furniture, such as parapets, etc.

Clause 8.3.1(6)

Wind pressures induced on the bridge from the passage of traffic are outside the scope of EN 1991-1-4.

Clause 8.3.1(7)

8.3.2. Force in x-direction – simplified method

Clause 8.3.2(1)

The simplified method applies when assessment of dynamic response is not necessary, so that the simplified x-force, F_w, is given by:

$$F_W = \tfrac{1}{2} \cdot \rho \cdot v_b^2 \cdot C \cdot A_{ref,x} \qquad (D8.1)$$

where C is the 'wind load factor', that corresponds to $C = c_e \cdot c_{f,x}$ the product of the exposure factor and the x-force coefficient. Table 8.2 gives a choice of four datum values of C, depending on the deck proportions b/d and the reference height z_e. Interpolation between these datum values is allowed, and is practically essential because the four datum values differ considerably and taking the largest may be grossly conservative.

UK NA 2.48

The NA is allowed to define alternative values of C to replace Table 8.4. The UK NA takes advantage of this choice, giving values in Table NA7 that are around 20% larger to comply with current UK practice.

8.3.3. Wind forces on bridge decks in z-direction

Clause 8.3.3(1)
UK NA 2.49

The force in the z-direction is normally referred to as the 'lift on the deck'. Values of $c_{f,z}$ may be given in the National Annex. Otherwise, Note 1 gives a recommended single value of $c_{f,z} = \pm 0.9$ (i.e. acting upwards *or* downwards), or alternatively less conservative values may be obtained from Fig. 8.6. The UK NA adopts the recommended value.

Clause 8.3.3(2)–(4)

The reference area $A_{ref,z}$ is the plan area of the deck, no end-effect factor is appropriate and the reference height z_e is the same as for the x-forces.

Clause 8.3.3(5)

Unless '*otherwise specified*' (by the 'appropriate authority') the z-force should be assumed to act on the deck eccentrically in the x-direction by a distance $e = b/4$ from the centre of the deck.

8.3.4. Wind forces on bridge decks in y-direction

Clause 8.3.4(1)

The force in the y-direction is the force along the axis of the deck caused principally by friction from wind blowing along the long deck. Values may be given in the National Annex, otherwise 25% of the forces in the x-direction are recommended for plated bridges (bridges with solid plate girders or box-girders) and 50% for truss bridges. This force is not usually significant in the design of bridges due to the small values of force and the high structural resistance along the axis of the deck. It may, however, control the design of restraints at expansion joints.

UK NA 2.50

The UK NA requires the longitudinal wind load to be taken as the more severe of:

- the longitudinal wind load on the superstructure of the bridge acting alone, or
- the nominal wind load in combination with the nominal live load as defined in equations (NA7) to (NA9).

Further rules are given to determine the wind loads on parapets and safety fences, brackets extending outside the main girder and trusses in skew winds (nominally 45° from the axis) and on the bridge piers.

8.4. Bridge piers

8.4.1. Wind directions and design situations

Clause 8.4.1(1)–(2)

The designer is required to identify the most unfavourable direction for the overall wind loads on the whole bridge structure for each load effect under consideration, and make separate calculations for all transient design situations during construction. This is needed because the most critical loads on piers usually occur as out-of-balance loads during assembly of the bridge.

8.4.2. Wind effects on piers

Clause 8.4.2(1)
Notes 1 and 2

No individual provisions are given in the EN for determining the wind loads on bridge piers. The designer is directed to the general provisions of clauses 7.6, 7.8 or 7.9.2 for structural sections and cylinders. The National Annex may give simplified rules for wind loads on bridge piers and/or procedures for assessing asymmetric loading. The recommended procedure is to remove the design wind load completely from all parts of the structure where the wind load is beneficial to the load effect under consideration.

The UK NA corrects this deficiency by including force coefficients for piers C_{fp} in Table NA8 that have been derived from Table 9 of BD 37/01[24] and by applying the asymmetry rules of UK NA 2.22.

UK NA 2.51 and 2.52

Additionally, the UK NA exploits the scope of Note 1 to introduce a simplified quasi-static procedure for in-line wind effects that applies to continuous bridges, an important exceptional case where 'unfavourable' loading is particularly important. Unfavourable loading does not necessarily involve complete spans being loaded.

UK NA 2.53

The procedure first identifies single-span bridges as suitable for the simplified procedure of clause 8.3.2, as modified by the UK NA, since unfavourable loads are not a critical design issue for this case. However, for continuous bridges, the procedure identifies:

- elements where the wind increases the load effect – for which the standard procedure in clause 5.3(2) is appropriate, and
- elements where the wind relieves the load effect – where the relieving force may be determined from expression (NA11) (using an overall force coefficient) or Expression (NA12) (using vector summation).

In the UK, allowance for the non-simultaneous action of peak pressures over the loaded length being considered may be provided through clause 5.3(2) using the value of the size factor given in Table NA3 and taking $b + h$ as the base length of the adverse area under wind loading. Application of the UK NA simplified procedure is qualified by three Notes:

- Note 1 gives a methodology for determining the critical adverse area for bridges of continuous construction.
- Note 2 introduces a requirement for design of bridges located close to the top of a hill, ridge, cliff or escarpment to '*allow for the significance of the orographic feature*'. Use of the size factors in Table NA3 is not permitted in this instance. Effectively, this Note requires the user to seek specialist advice for bridges in hilly terrain.
- Note 3 allows the 'division by parts' rule of Fig. 7.4 to be used for tall vertical elements such as piers and towers.

CHAPTER 9

Annexes

This chapter is concerned with the six Annexes to EN 1991-1-4. The material in this chapter is covered in the following clauses:

- General comments
- Terrain effects · *Annex A*
- Procedure 1 for determining the structural factor $c_s c_d$ · · · · *Annex B*
- Procedure 2 for determining the structural factor $c_s c_d$ · · · · *Annex C*
- $c_s c_d$ values for different types of structure · · · · · · · · · · · *Annex D*
- Vortex shedding and aeroelastic instabilities · · · · · · · · · · · *Annex E*
- Dynamic characteristics of structures · · · · · · · · · · · · · · · · *Annex F*

9.1. General comments

The status of the annexes to EN 1991-1-4 is that each annex is **informative**, unless its status is changed by the National Annex, which may declare that the annex is **normative** or that it should not be used at all. Unless the annex specifically provides for modification by the NA, e.g. as in the Note to clause A.2(1), the NA may not make changes to an annex. Instead it must adopt the annex as it is, or reject it in its entirety and provide a reference to a completely new annex which has been published separately as '*non-contradictory complementary information*' (NCCI). Any new annex can, of course, replicate any or most of the original annex, adding or changing parts as required. However, the CEN drafting rules prevent any new annex from being incorporated directly into the National Annex – which would be the most logical and convenient place for it to appear.

9.2. Annex A Terrain effects

The UK National Annex adopts Annex A as an informative annex, but makes changes and additions at locations of permitted National Choice or Nationally Determined Values within the annex.

UK NA 3.1

9.2.1. A.1 Illustrations of the upper roughness of each terrain category

This section provides illustrations of the '*upper roughness*' for each of the five defined terrain categories. The term '*upper roughness*' implies that terrain of intermediate roughness shall be placed under the next rougher category, i.e. terrain rougher than the upper roughness of Category 0, but less rough than the upper roughness Category I, classifies as Category I. This is not a generally conservative assumption since it always provides a higher roughness, and hence more shelter, than is appropriate. It would have been more appropriate to have illustrated the 'lowest permitted roughness' of each terrain category.

Clause A.1

UK NA 2.54
UK NA 3.1.1
There are no provisions in EN 1991-1-4 for altering the definitions of these standard terrain categories in the National Annex. However, the following clause A.2 allows the NA to define rules for transition effects. The UK NA exploits this opportunity to reduce the effective number of categories from five to three by defining transition rules such that the rural terrain Categories I and II are treated as a single category designated 'Country' and the urban terrain Categories III and IV are treated as the single category designated 'Town'. The remaining Category 0 is designated 'Sea'. The EN is not intended for off-shore structures and should not be used for that purpose. Category 0, 'Sea', is used for the sole purpose of defining the coastal boundaries of the other categories.

9.2.2. A.2 Transition between roughness Categories 0, I, II, III and IV

Clause A..2(1)
Transition between categories must be considered when calculating the peak velocity pressure q_p and the structure factor $c_s c_d$ and, by implication, the separate size factor c_s and dynamic factor c_d.

The procedure for transition between categories may be given in the National Annex, but two recommended procedures are provided in the EN to account for smooth-to-rough changes:

- Procedure 1 is the simpler. Following a transition to a rougher terrain, the upwind smoother terrain should be adopted when this is less than 2 km upwind when Category 0, or less than 1 km upwind when Category I, II or III. This means that sites less than 2 km away from a coastline should be treated as Category 0 for all sites, rural or urban. Otherwise, all suburban sites less than 1 km from the rural–suburban boundary will be treated as rural, and all city-centre sites less than 1 km from the suburban–urban boundary will be treated as suburban.
- Procedure 2 is the more accurate. It defines a threshold height above ground, below which the new rougher category may be used and above which the previous smoother category must be used. This threshold height increases with distance, or 'fetch', down-wind of the roughness change. While more accurate than Procedure 1, Procedure 2 assumes that the wind speed is instantly in equilibrium following each change in rough-ness whereas, in reality, the wind speed changes gradually with height above ground and

UK NA 3.1.1
distance from the roughness change (fetch). The transition procedure in the UK NA implements these gradual changes and is effectively a 'Procedure 3'.

When the displacement height h_{dis} is used (see section 4.3.2 above), these procedures will be very conservative if the standard displacement height for the effective terrain is used, but may not be conservative if the actual displacement height is used.

There are no recommended procedures for the case of a rough-to-smooth transition. The new smoother terrain category is deemed to apply immediately. This is a conservative assumption since no account is taken of the shelter provided by the previous rougher terrain.

UK NA 2.54
In nature, the ground roughness varies continuously and often lies somewhere between the datum categories. However, the change from 'Sea' to 'Country' (or coastal town) is always abrupt as is, typically, the change from 'Country' to 'Town'. This latter case stems from the economic imperatives that drive urban development: the building density of an urban area does not increase gradually, but tends to occur in phases directly to the final building density. For these reasons, these three categories – 'Sea' for 0, 'Country' for I and II and 'Town' for III and IV – are easily recognized on the ground or in standard mapping,

UK NA 3.1
which is why BS 6399-2[1] uses these descriptions. The UK NA effectively adopts the same rules, as described earlier in sections 4.3.2, 4.4 and 4.5. Accordingly, the UK NA does not permit the use of Annex A2.

9.2.3. A.3 Numerical calculation of orography coefficients

Clause A.3(1)–(4)
The orography coefficient c_o is a factor that applies to the mean wind speed. The method for calculating the orography coefficient c_o may be defined in the National Annex, otherwise the recommended method can be used. The UK NA adopts the recommended method.

The recommended method for calculating the speed-up of the mean wind speed over isolated hills, ridges and cliffs is two-dimensional and does not predict the speed-up around the flanks of hills or from funnelling in steep-sided valleys. Key parameters of the orography are the upwind slope and the position of the site relative to the crest.

The orography model assumes that only the mean wind speed is affected, speeding up or slowing down, while rms value of the turbulence remains unaffected. This has the result that the turbulence intensity (rms turbulence/mean wind speed) decreases in value proportionally to the increase in the mean wind speed. This is why the orography coefficient c_o appears in the denominator of the equation for turbulence intensity $I_v(z)$, equations (D4.8a) and (D4.8b).

Application of the orography method requires the user to determine the cross-sectional shape of the orographic feature along a line through the site in the wind direction, then to represent the actual shape by an appropriate triangular shape. This is probably the most time-consuming part of the code, but also the most important because the speed-up over orography can be the largest single factor in the calculation of velocity pressure.

This method is virtually identical to the rules in BS 6399-2,[1] so users may refer to the respective guidance[20] for practical examples of its use. This guidance is particularly helpful when the cross-sectional shape of the hill is not closely triangular.

9.2.4. A.4 Neighbouring structures

Buildings that are significantly taller than their neighbours tend to deflect high wind speeds down to ground level, which will increase the wind loads on the neighbours. Most codes, including BS 6399-2,[1] warn of this effect, but give no guidance other than to seek expert advice. An exception to this is the ECCS model code[12] which is the source of the procedure adopted in EN 1991-1-4. This defines a new reference height above ground z_n, for use when calculating the increased wind loads on the neighbours ($z_e = z_n$). These rules are entirely empirical. A better estimate of the effect of neighbouring structures that includes possible shelter would be obtained by wind tunnel testing. The UK NA adopts this part of the annex without changes.

Clause A.4(1)

9.2.5. A.5 Displacement height

Where buildings, or other obstructions, are closely spaced the wind is displaced upwards by h_{dis} to form an effective ground-plane just below the average height of the roof tops h_{ave}. The rule given in the annex refers to Category IV terrain, but there is no technical reason why this would not also apply to Category III terrain. Indeed, the UK NAD uses common rules for Categories III and IV which include h_{dis}. The procedure is identical to the rule in BS 6399-2,[2] whereby $h_{dis} = 0.8h_{ave}$ for a distance up to $2h_{ave}$ downwind of the nearest building, tapering linearly to $h_{dis} = 0$ at a distance of $6h_{ave}$ downwind.

Clause A.5(1)

This procedure is conservative – it is known that some shelter still exists $20h_{ave}$ downwind – but some conservatism is required to allow for the possibility of demolition of a neighbour. In effect, the procedure allows for the displacement caused by the massing of upwind buildings but neglects the direct shelter from the wake of any individual building.

The procedures in A.5 comply with UK practice and the UK NA adopts them without changes.

9.3. Annex B Procedure I for determining the structural factor $c_s c_d$

This is the first of two alternative recommended procedures. The National Annex is required to declare which may be used. The UK NA has opted to separate the structural factor into size factor c_s and dynamic factor c_d, but has used this procedure to derive their values. However, the UK NA also allows direct use of Annex B, but see the comments earlier in section 6.1.

Clauses B.1–B.4
UK NA 3.2

Procedure 1 is a 'euro-implementation' of the classic Davenport methodology, which is fully described in many wind engineering texts. The method is in widespread use around the world, principally for vertical cantilever structures such as tall buildings, and has stood the test of time. The response is divided into a random wide-band 'background' part, B, caused by turbulent buffeting and a narrow-band 'resonant' part, R, caused by dynamic amplification at the lowest natural frequency of the structure. Background and resonant parts are estimated separately, then are combined to give an overall response. Methods are included to determine the number of load cycles for use in fatigue estimates and to estimate service displacements and accelerations.

The procedure of Annex B is implemented in a number of steps:

- The turbulence length scale $L(z)$ is determined from Expression (B.1). This depends on the height above ground, z, and the terrain roughness length z_o. The turbulence length scale describes the characteristic size of the dominant turbulent eddies.
- The turbulence spectrum, expressing the distribution of turbulent energy over frequencies, is determined from $L(z)$ using Expression (B.2). This is used to obtain the resonant part R at the fundamental natural frequency of the structure.
- The background factor B^2, which allows for the lack of correlation of pressure over the structure, is determined from $L(z)$ and the structural dimensions b and h.
- The up-crossing frequency, ν is obtained from Expression (B.5).
- The peak factor k_p, which is the ratio of the maximum fluctuating response to the rms value, is determined from Expression (B.4), using ν.
- The resonant response function R^2, which allows for turbulence in resonance with the structure, is determined from Expression (B.6).
- The derived values of B^2, R^2 and ν are used in Expressions (6.1), (6.2) and (6.3) to give the combined structural factor $c_s c_d$ and the separate size factor c_s and dynamic factor c_d.

Annex B also gives a method to estimate the number of times that a load effect level is exceeded for use in fatigue life calculations and a method to estimate the service displacements and accelerations of a vertical structure.

9.4. Annex C Procedure 2 for determining the structural factor $c_s c_d$

Clauses C.1–C.4
UK NA 3.3

This is the second of two alternative recommended procedures. The National Annex is required to declare which may be used. The UK NA does not permit this procedure to be used in the UK.

Procedure 2 is a new and virtually untested method which, it is claimed, gives a result within 5% of Procedure 1. The only virtue of this procedure appears to be marginally simpler implementation. The method is derived from Dyrbye and Hansen,[14] and the reader is directed there for further commentary.

9.5. Annex D $c_s c_d$ values for different types of structure

Clause D(1)

This annex presents a series of design charts giving values of $c_s c_d$ for a range of common structures. While these values may indeed be typical – and this depends on the assumed values for the controlling parameters which are listed alongside each graph – the crucial requirement to be able to identify unexpected atypical behaviour is circumvented if these charts are used directly. These charts should be used to obtain initial 'ballpark' values or to check whether values calculated using the full methods of Annex C or D are very different from typical. A small difference may indicate that the structure is not typical, but a large difference may indicate a calculation error.

UK NA 3.4

The UK NA does not permit Annex D to be used in the UK.

9.6. Annex E Vortex shedding and aeroelastic instabilities

This annex gives a simplified overview of a highly complex set of dynamic instabilities that may affect certain structures. These instabilities are most relevant for flexible 'line-like' structures, i.e. structures that are long and thin. These include large-span bridges and tall, slender chimneys, where the design rules are framed to exclude the possibility of instability. Instabilities are most unlikely to occur with building structures. The procedures in Annex E are complex, difficult to apply and not necessarily sufficient to exclude all significant risk of instability. Accordingly, the UK NA does not permit Annex E to be used in the UK and refers to the background document[xx], which gives a replacement annex. In any case, the designer should seek expert advice for structures that may be susceptible to these effects.

Clauses E.1–E.4
UK NA
2.56–2.63,
UK NA 3.5

9.6.1. E.1 Vortex shedding

The introductory 'General' paragraph correctly describes vortex shedding as occurring '*when vortices are shed alternately from opposite sides of a structure*'. This is caused by a quasi-periodic interaction between the two shear layers separating from either side of the structure. Vortex shedding cannot occur if there is only one shear layer, e.g. from the top edge of a long free-standing wall, and is suppressed when the structure is slightly porous (~8% porosity, or greater) because the flow through the structure interferes with the interaction between the shear layers. A lattice structure will not shed vortices associated with its overall size, but will experience vortex shedding of individual long members. As the excitation occurs over a range of frequencies, the structure will respond quasi-statically over this range except when the structure's fundamental natural frequency falls within this band. In which case, the vortex shedding locks on to the resonant motion of the structure and becomes periodic, leading to larger response amplitudes – this is called 'lock on'.

Clause E.1.2 gives several criteria to assess whether vortex shedding needs to be considered:

(1) The structure must be long and slender, such that the length l is at least 6 times the cross-wind breadth b. To this should also be added that the structure should not be more than 8% porous.
(2) Next, the critical velocity for vortex shedding, v_{crit}, must lie within the expected range of wind speed for the structure. As vortex shedding can occur in the larger gusts, the maximum critical velocity is greater than the design mean wind speed, v_m. The recommendation is that vortex shedding should be investigated when $v_{crit} > 1.25 \cdot v_m$ (Expression (E.1)).
(3) The critical velocity for vortex shedding, v_{crit}, is determined from the crosswind breadth of the structure, b, the natural frequency of vibration, n, and the Strouhal number, St, associated with the cross-sectional shape. Table E.1 gives values of St for various common cross-sections.
(4) The ability of the structure to absorb and dissipate the energy from vortex shedding depends on the structural damping, and this is expressed by the Scruton number, Sc. Expression E.4 giving Sc requires values for the logarithmic decrement of the structural damping, δ, and the equivalent mass-per-unit-length of the structure in the mode of vibration, m_e.

It is at step 4, above, that the process of assessing vortex shedding gets rather complicated. If the user does not have an understanding of the aerodynamic and structural characteristics and does not have some experience of the calculation process, the subsequent estimation of motion amplitudes and accelerations is likely to be difficult and prone to error.

9.6.2. E.2 Galloping

Classical galloping is an aeroelastic instability of long slender structures that occurs at all wind velocities above a critical threshold value. This critical velocity depends on Scruton number, Sc, crosswind breadth, b, natural frequency, n, and a factor a_G which, in turn, depends on the cross-sectional shape and is given by Expression (E.21). Only some particular cross-sections are able to gallop, and then only for some particular wind angles. Many sharp-edged sections will gallop, including typical bridge decks. Circular sections cannot gallop (classically) unless their cross-section is modified by ice accretion or rivulets of rain. Rain-induced galloping is a particular problem for the primary support cables of cable-stayed bridges.

9.6.3. E.3 Interference galloping

Interference galloping occurs when two or more sections are arranged close enough together such that one may move in and out of the turbulent wake cast by the other, giving it its alternative name of 'wake galloping'. Pairs of circular cylinders, e.g. power-line conductors, are prone to wake galloping, as are cylindrical elements lying parallel to other structural sections, e.g. pipes and cable-risers. The regular 'ting-ting-ting' noise frequently heard in yacht marinas is caused by rope halyards slapping against masts as interference galloping causes the halyards to move in and out of the mast wakes. With structural elements, fretting damage will occur if elements clash together, otherwise the principal problem is fatigue. The critical velocity for wake galloping depends on the same parameters as classical galloping, and is given for the case of two identical cylinders by Expression (E.23). There is not sufficient information available to frame rules for other section shapes and avoidance of the problem is best ensured by wind tunnel testing.

9.6.4. E.4 Divergence and flutter

These are instabilities that occur for flexible plate-like structures, including signboards and suspension bridge decks. Divergence is a static torsional instability that leads to sudden catastrophic failure unless the deflection is limited by aerodynamic stall, when it becomes cyclic 'stall-flutter' or 'stop-sign flutter'. Classical flutter is a torsion-translation coupled dynamic instability that builds in amplitude to a limit cycle or to failure.

For divergence or flutter to be likely, the lowest natural frequency must be in torsion, or at least the torsional frequency must not be greater than twice the lowest translational frequency. The calculation method for the critical wind velocity requires the aerodynamic derivatives for the cross-section. 'Aerodynamic derivatives' are the rates of change of the principal loading coefficients with wind angle. It is difficult enough to get reliable values of the principal loading coefficients from published sources: getting reliable derivatives is almost impossible without resorting to wind tunnel testing.

The simplest way to avoid the possibility of divergence or flutter is to provide adequate torsional stiffness and to ensure that the centre of lift force is coincident or downwind of the torsional centre. Otherwise, the stability of susceptible structures needs to be ascertained using wind tunnel testing, as is normal for suspension bridges.

9.7. Annex F Dynamic characteristics of structures
9.7.1. F.1 General

Clauses F.1–F.5
UK NA 3.6

This annex gives a simplified overview of structural dynamics which is not comprehensive enough to substitute for the many textbooks on the subject. It does, however, give recommended values of the structural damping logarithmic decrement δ of many common structural forms. The UK NA permits its use as an informative annex.

9.7.2. F.2 Fundamental frequency

The fundamental frequency, the lowest natural frequency of vibration of a structure, is frequently needed in the EN methods for dynamic response. This clause provides some simple estimators for the fundamental natural frequency.

- For cantilevers

$$n_t \approx \frac{9.81}{2 \cdot \pi \cdot \sqrt{x_t}} \text{ Hz, Expression (F.1)}$$

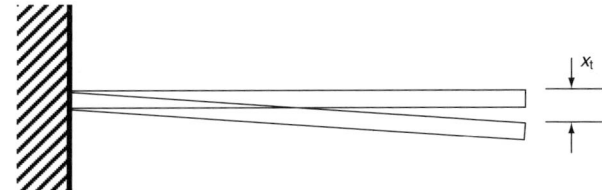

Fig. D9.1. Deflection of cantilever under self-weight

where x_t (in metres) is the deflection at the tip of the cantilever that would occur under self-weight if supported horizontally, as illustrated in Fig. D9.1.

- For tall buildings $n_t = 46/h$ Hz, where h is the height in metres.
- For chimney stacks, using Expressions (F.3) and (F.4).
- For ovalling of shell structures, using Expression (F.5).
- For bending of plate or box-girder bridges, using Expression (F.6).
- For torsion of box-girder bridges, using Expressions (F.7) to (F.12).

9.7.3. F.3 Fundamental mode shape

Annex F represents the shape of the fundamental mode of structures cantilevered from the ground by a power law, with an index varying from 0.6 for framed structures to 2.5 for lattice towers, as indicated in Fig. D9.2.

Annex F represents the fundamental mode shapes of simply-supported or clamped-end structures, such as bridge decks, as shown in Fig. D9.3.

9.7.4. F.4 Equivalent mass

Expression F.14 may be used to estimate the equivalent mass per unit length m_e of the fundamental mode, reproduced here as equation

$$m_e = \frac{\int_0^t m(s) \cdot \phi^2(s) \cdot \mathrm{d}s}{\int_0^t \phi^2(s) \cdot \mathrm{d}s} \qquad (D9.1)$$

Equation (D9.1) is the exact universal expression. A simple estimate for cantilever structures is the average mass per unit length over the upper third of the structure. A simple estimate for

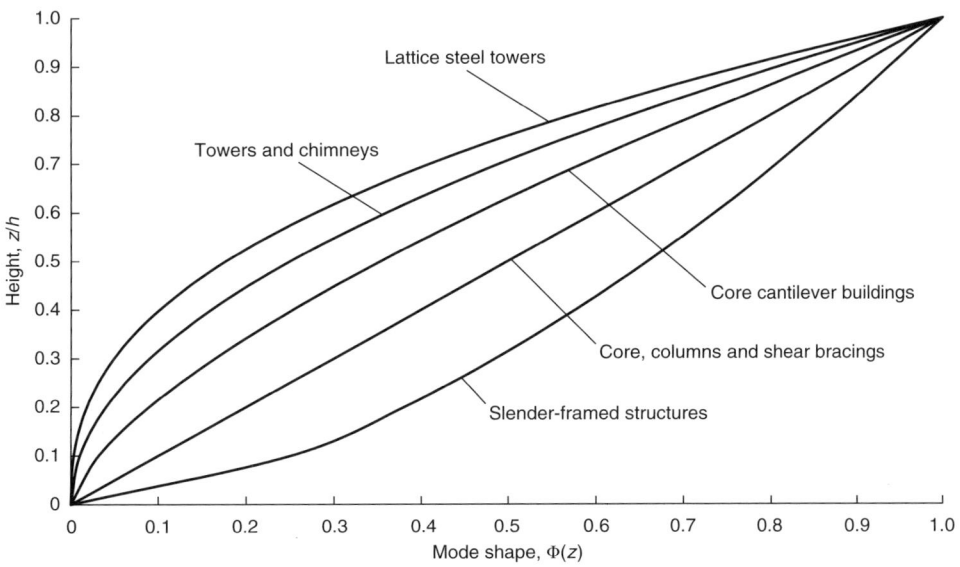

Fig. D9.2. Fundamental mode shapes for structures cantilevered from the ground

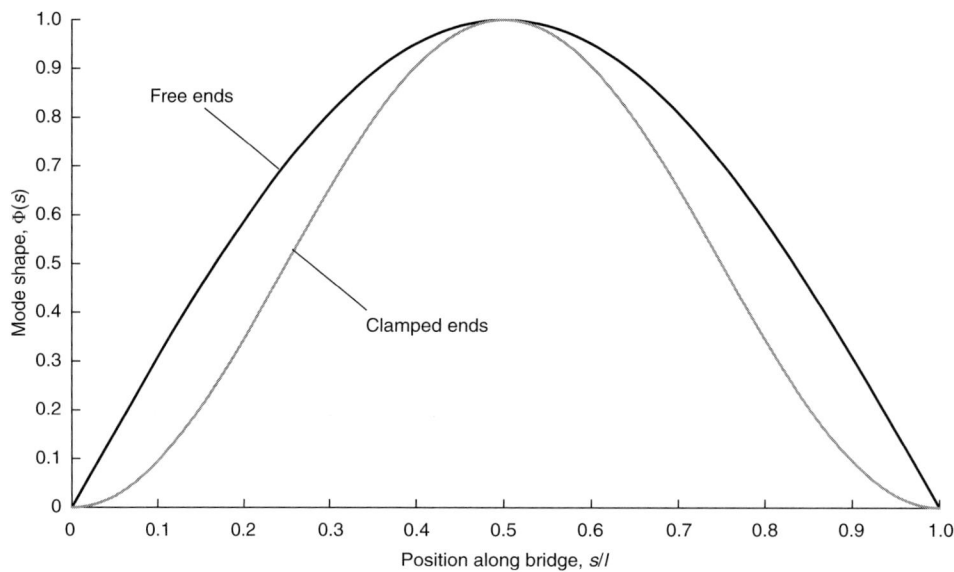

Fig. D9.3. Fundamental mode shapes for simply-supported and clamped structures

long structures simply supported at either end, e.g. bridges, is the average mass per unit length over the middle third of the structure.

9.7.5. F.4 Logarithmic decrement of damping

The logarithmic decrement of damping is the sum of the logarithmic decrement of all sources of damping: structural damping, aerodynamic damping and the damping of any passive or active special devices attached to the structure. Values of structural damping for many forms of structure are given in Table F.2. Aerodynamic damping for the in-wind vibration of cantilever structures is given by Expressions (F.16) and (F.17).

Postscript

The reader will have noted this author's criticisms of various parts of EN 1991-1-4, from its structure, through its content to its omissions. The 'vertical' structure of the Eurocodes, where the content is sliced up strictly by topic, leads to the situation where the design of the simplest of structures may require five or more ENs in place of a single national code specific to that structure. This author favours the 'horizontal' structure where the code relates to the structure (e.g. buildings, bridges, lattice towers) and can focus on the particular needs of that structure, eliminating irrelevant data and methods.

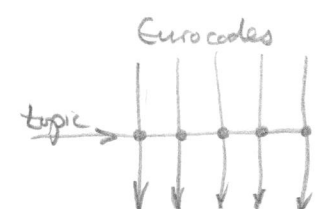

The task of 'harmonizing' rules for wind loading of structures across Europe is extremely difficult, if not impossible. It is therefore not surprising that this first implementation of EN 1991-1-4 has many flaws, omissions and contradictions. The EN acknowledges many of these by the large number of instances of National Choice that are not just limited to National Values (NVs), but also extend to replacement of recommended rules. Ideally, the role of the National Annex should be limited to providing NVs that are particular to the wind climate of the Member State, i.e. that are geographically dependent. The EN invites changes that extend the NA well beyond this role, which is the inevitable result of the failure to agree a common approach and consistent values. The NAs are due to be amalgamated into the EN at the end of the 'coexistence period' but, given the extent of disagreement across Member States, this task will be more difficult to achieve than the progress made so far.

The use of codes of practice differs considerably across the Member States, as does their relationship to laws and regulations governing safety. At the one extreme, designs may be produced from first principles without reference to codes of practice at all. During development of the initial draft, one member of the drafting panel proposed that the EN rules for bridges should contain only Principles – with no implementation rules or data – because 'all bridges in Switzerland are designed by competent engineers'. At the other extreme, codes of practice may be referred to directly or indirectly in laws or statutes, so that use of the code of practice becomes mandatory in design. The UK situation is closer to this second extreme.

The structure and format of EN 1991-1-4 is complicated and obscure to the degree that it is very difficult to apply it to a particular design without a strong likelihood of error. Many of the recommended implementation rules are unnecessarily complex and, where simplifications have been attempted, the resulting loss of accuracy outweighs any benefit. The UK National Annex has taken every opportunity afforded by EN 1991-1-4 to correct the inherent errors while also making implementation simpler and less prone to error.

Errors in the EN implementation include the following:

- The EN implementation of size and dynamic effects applies only to the overall structure. Loads on elements of buildings are conservative if the EN '1 m^2' and '10 m^2' coefficients are used as recommended.

- Non-conservative error in linearizing the gust factor model.
- Torsional effects are non-conservative because orthogonal loading coefficients are symmetrical and the recommended rule for buildings is not an adequate substitute.
- Overall drag obtained from summation of orthogonal pressures is conservative because maximum front-face loads occur at different wind angle from maximum rear-face loads.
- The frictional component of overall loads is not mitigated for size effect.
- 'Open-sided' buildings, typically grandstands, aircraft hangars and other special industrial building forms are not covered by the internal pressure rules.
- Loads on permanently unclad lattice structures may be grossly conservative and exceed the loads for a fully clad structure because self-shelter is not addressed.
- Elective dominant openings are specifically referenced as 'accidental' conditions and not as service conditions. However, service conditions are implied in the 'catch-all' Principle, clause 7.2.9(1)P.

Many of these have been corrected by the UK National Annex, including the following:

- Separate size-effect factor made applicable to elements of buildings removes conservative error.
- Full gust factor model removes non-conservative error.
- The reduced upwind roughness for sites close to large inland lakes should be accounted for by treating lakes >1 km across as 'Sea' – Category 0 instead of Category I.
- Allowance for torsion on buildings is increased to eliminate non-conservative error. This should also be used for piers of bridges.

However, restrictions on the scope of the NA prevent the correction of some errors. This Guide offers advice to deal with these cases, including the following:

- 'Open-sided' buildings should be addressed using NCCI.
- Permanently unclad lattice structures should be assessed using NCCI.
- Elective dominant openings should be treated as a serviceability condition, with appropriate probability and seasonal factors as well as the corresponding partial load factor.

Other instances where the EN does not permit patently obvious errors to be corrected will need to be addressed by the revision at the end of the 'coexistence period'. In the meantime, experience of previous 'coexistence', such as the five years that CP3:ChV:Pt2 coexisted with BS 6399-2 in the UK, indicates EN 1991-1-4 is unlikely to be taken up widely until this is forced by withdrawal of the national codes, driven by the significant costs of implementing the change which will fall as an overhead charge to most design firms. Thus flaws in the EN will not become evident to the majority of users until it is too late for them to influence the revision.

References

1. British Standards Institution (1997) *Loading for Buildings. Part 2: Code of Practice for Wind Loads*. BSI, London, BS 6399-2.
2. Gulvanessian, H., Calgaro, J.-A. and Holický, M. (2000) *Designers' Guide to EN 1990. Eurocode: Basis of Structural Design*. Thomas Telford, London.
3. European Commission (2003) *Guidance Paper L (Concerning the Constructions Products Directive – 89/106/EEC, Application and Use of Eurocodes*. EC, Brussels, ENTR/G5.
4. International Organization for Standardization (1998) *General Principles on Reliability of Structures*. ISO, Geneva, ISO 2394.
5. International Organization for Standardization (1999) *Bases for Design of Structures – Notations – General Symbols*. ISO, Geneva, ISO 3898.
6. International Organization for Standardization (1987) *General Principles on Reliability of Structures – List of Equivalent Terms*. ISO, Geneva, ISO 8930.
7. International Organization for Standardization (2000) *Quality Management and Quality Assurance – Vocabulary*. ISO, Geneva, ISO 8402.
8. Building Research Establishment (1999) *Wind Loading on Buildings, Parts 1–3*. Digest 436. BRE, Watford.
9. Davenport, A. G. (1961) The application of statistical concepts to the wind loading of structures. *Proceedings of the Institution of Civil Engineers*, **19**, 449–471.
10. Cook, N. J. and Mayne, J. R. (1980) A refined working approach to the assessment of wind loads for equivalent static design. *Journal of Wind Engineering and Industrial Aerodynamics*, **6**, 125–137.
11. Cook, N. J. (1982) Further development of a working approach to the assessment of wind loads for equivalent static design. *Journal of Wind Engineering and Industrial Aerodynamics*, **9**, 389–392.
12. European Convention for Constructional Steelwork (1987) *Recommendations for Calculating the Effects of Wind on Constructions*. Technical Committee 12 – Wind, Report 52, 2nd edn. ECCS, Brussels.
13. Jackson, P. S. and Hunt, J. C. R. (1975) Turbulent wind flow over a low hill. *Quarterly Journal of the Royal Meteorological Society*, **101**, 929–955.
14. Dyrbye, C. and Hansen, S. O. (1997) *Wind Loads on Structures*. Wiley, London.
15. Cook, N. J. (1990) *The Designers' Guide to Wind Loading of Building Structures. Part 2: Static Structures*. Butterworths, London.
16. Newberry, C. W. and Eaton, K. J. (1974) *The Wind Loading Handbook*. HMSO, London.
17. Building Research Establishment (1985) *BRE Digest 295: Stability under Wind Load of Loose-laid Insulation Boards*. BRE, Watford.
18. British Standards Institution (2003) *British Standard Code of Practice for Slating and Tiling. Part 1 – Design*. BSI, London, BS 5534.

19. British Standards Institution (2005) *Design of Masonry Structures: Common Rules for Reinforced and Unreinforced Masonry Structures.* BSI, London, BS EN 1996-1-1:2005.

20. Cook, N. J. (1999) *Wind Loading – A Practical Guide to BS 6399-2, Wind Loads on Buildings.* Thomas Telford, London.

21. British Standards Institution (In production) *Design of Steel Structures: Towers, Masts and Chimneys, Part 3.1: Towers and Masts.* BSI, London, prEN 1993-3-1.

22. British Standards Institution (2004) *Scaffolds – Performance Requirements and General Design.* BSI, London, BS EN 12811:2003.

23. Highways Agency (2001) *Design Manual for Roads and Bridges. Vol. 1, Highway Structures: Approval Procedures and General Design. Section 3: General Design. Part 3: Design Rules for Aerodynamic Effects on Bridges.* HMSO, London, BD 49/01.

24. Highways Agency (2001) *Design Manual for Roads and Bridges, Volume 1, Section 3: General Design: Loads for Highway Bridges.* HMSO, London, BD 37/01.

xx. British Standards Institution (2007) *PD 6688, Background Document to National Annex to BS EN 1991-1-4* (in preparation).

yy. ASCE (1999) *Manuals and Reports on Engineering Practice, No. 67. Wind Tunnel Studies of Buildings and Structures.*

Index

Abbreviations: NA – National Annex; UK NA – United Kingdom National Annex. Page numbers in italics refer to diagrams and illustrations.

inside front cover – Eurocode Designer's Guide Series listing

1. First version limited to (a) buildings <200m (b) bridges <200m

2. Guyed masts and <u>lattice towers</u> covered separately

   EN 1993-3-1
   (Design of Steel Structures, Part 3.1, Towers & Masts, 2006 [to be published])

5. Design assisted by tests and measurements

7. Symbols and notation

14. – Characteristic values (chance that characteristic value is   (GTT?)
    exceeded at least once in return period is approximately 64%)

∗ ———▷ – Characteristic annual risk of 0.02 of being exceeded in <u>each & every year</u>
    – Mean recurrence interval (return period) = 1/0.02 = 50 years

20. Probability factor

30. Summary of wind <u>velocity</u> parameters (EN vs BS)

33. Summary of wind <u>action</u> parameters (EN vs BS)

41. Size factor (1m² vs. 10m²)   (GTT?)

54. Lattice structures

64. Orography coefficients

67. Vortex shedding and aeroelastic instabilities

68. Dynamic characteristics of structures

71. Criticisms of EN (flaws, omissions, contradictions; complicated, obscure)

72. EN unlikely to be taken up widely during coexistence period

∗ but nb, in limit state codes, ultimate factored loads are frequently specified
as having a 5% probability of not being exceed in life of the structure.
(5% in 50 yrs = 5/100 × 1/50 = 1/1000, hence return period is 1000 years)